Jan Philipp Dapprich / Annika Schuster

Philosophy and Logic of Quantum Physics

An Investigation of the Metaphysical
and Logical Implications of Quantum Physics

Bibliographic Information published by the Deutsche Nationalbibliothek
The Deutsche Nationalbibliothek lists this publication in the Deutsche Nationalbibliografie; detailed bibliographic data is available in the internet at http://dnb.d-nb.de.

Library of Congress Cataloging-in-Publication Data
Names: Dapprich, Jan Philipp, 1992- | Schuster, Annika, 1989-
Title: Philosophy and logic of quantum physics : an investigation of the metaphysical and logical implications of quantum physics / Jan Philipp Dapprich, Annika Schuster.
Description: Frankfurt am Main : Peter Lang, 2016. | Series: Philosophical foundations of the sciences and their applications, ISSN 2191-3706 ; vol. 5 | Includes bibliographical references.
Identifiers: LCCN 2015037768 | ISBN 9783631667255
Subjects: LCSH: Quantum theory. | Physics--Philosophy.
Classification: LCC QC174.12 .D354 2016 | DDC 530.1201--dc23 LC record available at http://lccn.loc.gov/2015037768

The authors thank the Chair of Theoretical Philosophy of Heinrich-Heine-University Düsseldorf for their monetary support which made this publication possible.

ISSN 2191-3706
ISBN 978-3-631-66725-5 (Print)
E-ISBN 978-3-653-06286-1 (E-Book)
DOI 10.3726/978-3-653-06286-1

© Peter Lang GmbH
Internationaler Verlag der Wissenschaften
Frankfurt am Main 2016
All rights reserved.
Peter Lang Edition is an Imprint of Peter Lang GmbH.

Peter Lang – Frankfurt am Main · Bern · Bruxelles · New York · Oxford · Warszawa · Wien

All parts of this publication are protected by copyright. Any utilisation outside the strict limits of the copyright law, without the permission of the publisher, is forbidden and liable to prosecution. This applies in particular to reproductions, translations, microfilming, and storage and processing in electronic retrieval systems.

This publication has been peer reviewed.

www.peterlang.com

PHILOSOPHISCHE GRUNDLAGEN DER WISSENSCHAFTEN UND IHRER ANWENDUNGEN

PHILOSOPHICAL FOUNDATIONS OF THE SCIENCES AND THEIR APPLICATIONS

Herausgegeben von / Edited by Gerhard Schurz

BD./VOL. 5

Zu Qualitätssicherung und Peer Review der vorliegenden Publikation

Die Qualität der in dieser Reihe erscheinenden Arbeiten wird vor der Publikation durch den Herausgeber der Reihe geprüft.

Notes on the quality assurance and peer review of this publication

Prior to publication, the quality of the work published in this series is reviewed by the editor of the series.

Philosophy and Logic of Quantum Physics

Acknowledgements

Thanks goes to the many physicists and philosophers who have challenged my understanding of quantum physics and thus helped develop the views presented in this work. Particular thanks is due to Prof. Gerhard Schurz for making this publication possible.

Jan Philipp Dapprich

This text was written under the supervision of Univ. Prof. Dr. Gerhard Schurz and Univ. Prof. Dr. Manuel Bremer, whom I wish to thank for their great supervision and brilliant comments.

Thank you Sara, Nina, Simon, Stephie, Alex and Philipp for proof-reading and valuable discussions on the topic and Matthias, Thomas, Leon, Lisa, Berenice, Svenja, Jutta, my and Matthias' family for your continuous love and support.

Annika Schuster

Contents

Joint Foreword 11

Part 1: Theory and Observation in Quantum Physics

  1.   Introduction 15

Part I.   Theory and Observation 17
  2.   The Relationship between Theory and Observation 17
  3.   Classification of Terms and Sentences 22
      3.1 Empirical Terms 22
           3.1.1 Observational Terms and Ostensive Learnability .. 22
           3.1.2 Empirical Disposition Terms 24
      3.2 Theoretical Terms 24
           3.2.1 Theoretical Terms in the Broad Sense 24
           3.2.2 Theoretical Terms in the Narrow Sense 24
      3.3 Classification of Statements 25
  4.   Theory (In-)dependence 26
      4.1 Theory (In-)dependence of Observational Language 26
      4.2 Theory (In-)dependence of Observation 28
      4.3 Linguistic Relativism 30
  5.   Ontological Status of Scientific Theories 32
      5.1 Realism and Antirealism 32
      5.2 Kinds of (Anti-)Realism 34
           5.2.1 Fundamental Positions 34
           5.2.2 Advanced Positions 36
      5.3 The Case against Solipsism 37
      5.4 The Case against Metaphysical Realism 38

	5.5	The Case against Radical Constructivism 39
	5.6	The Case for and against Constructive Realism 39
		5.6.1 Inference of the Best Explanation 39
		5.6.2 Pessimistic Meta-Induction 40
		5.6.3 Intertheoretical Correspondence 40
		5.6.4 Justification of Abduction according to Schurz 41
		5.6.5 The Logical Positivist Objection 42
	5.7	Towards a Modern Radical Empiricism 44
		5.7.1 The Independence of Theory 44
		5.7.2 Observational Language as Ostensively Learnable Language 44
		5.7.3 The Limited Theory-Ladeness of Observational Language 45
		5.7.4 Literalness of Theories 45

Part II. Basics of Quantum Physics 47

6. The Classical Picture of the World 47
7. Experimental Basis of Quantum Physics 50
8. The Quantum Explanation 53
   - 8.1 Wave-Particle Duality 54
   - 8.2 Schrödinger Equation and Born Rule 55
   - 8.3 Observables 56
   - 8.4 Einstein-Podolski-Rosen Paradox and Quantum Entanglement 56
9. No-go Theorems 57
   - 9.1 Bell's Theorem 58
   - 9.2 Kochen-Specker Theorem 58
10. Interpretations 61
    - 10.1 Schrödinger's cat 61
    - 10.2 Subjectivist Interpretations 62
    - 10.3 Objective-Collapse Interpretations 62

    10.4 Modal Interpretations 63
    10.5 Many-Worlds Interpretation 63
    10.6 Hidden-Variable Interpretations 63
    10.7 Operational Approach 64

Part III. Reflections on Quantum Physics 65

11. Theory and Observation in Quantum Physics 65
    11.1 Observables and Observation 65
    11.2 Theory or Interpretation? 66
    11.3 Theory and Ontology 68
    11.4 Structural Correspondences of Interpretations 69
12. Is Quantum Physics Acceptable? 69
    12.1 Indeterminism 70
    12.2 Value Definiteness 71
    12.3 Measurement Problem 73
13. Ontology of Quantum Physics 74
    13.1 Value Definiteness and Ontology 74
    13.2 Interpretations and Ontology 76
        13.2.1 Subjectivist Interpretations 76
        13.2.2 Objective-Collapse Interpretations 77
        13.2.3 Modal Interpretations 78
        13.2.4 Many-Worlds Interpretation 78
        13.2.5 Hidden-Variable Interpretations 79
        13.2.6 Operational Approach 79
        13.2.7 Interpretations and Ontological Parsimony 80
    13.3 Literalness and the Challenge of Quantum Gravity 81
14. Conclusion 82

References 85

Part 2: Quantum Logic

Introduction 93

1. **Fundamentals of Classical Logic and Quantum Mechanics** 95
   1.1 Peculiarities of Quantum Mechanics 95
   1.2 Classical Logic 98

2. **Quantum logic and the distributive law** 103
   2.1 Birkhoff and von Neumann's argument 104
       2.1.1 Popper's criticism 106
       2.1.2 Schurz's criticism 108
   2.2 Putnams's argument 111
       2.2.1 Dummett's criticism 114
       2.2.2 Stachel's criticism 115
   2.3 Conclusion on the distributive law 117

3. **Meaning and bivalence in Putnam's quantum logical system** 119
   3.1 Invariance of meaning 120
   3.2 Bivalence 124

4. **Conclusion** 129

References 131

Joint Conclusion 133

Joint Foreword

(by Jan Philipp Dapprich and Annika Schuster)

The evidence and background theory of quantum mechanics have raised questions among both physicists and philosophers resulting on the one hand in several interpretations of the theory and on the other hand in logical systems which are to reflect its unique structure. Its radical consequences such as indeterminism with regard to measurable, but incompatible, quantities and value indefiniteness provided philosophers with empirical evidence for arguments against realism and classical logic. In the two parts of this book we provide an overview over the interpretations of quantum mechanics as well as some of the quantum logical arguments and investigate the implications of quantum mechanics on ontology and classical logic.

The book is designed for readers without background knowledge in neither quantum mechanics nor logic and ontology. Each part begins with an introduction covering the phenomena needed for our philosophical analysis. But it surely would go beyond the scope of this book to review all the details in all the disciplines, which is why it will probably be more accessible with some foundations in these areas.

The two parts of this book can be read independent from each other as they are based on our Master's resp. Bachelor's thesis which we handed in at the Heinrich-Heine-University Düsseldorf.

The first part by Jan Philipp Dapprich is called "Theory and Observation in Quantum Physics". It is divided into three parts. The first part gives an introduction to the relationship between theory and observation and proposes a classification of terms and sentences. Then it discusses the theory dependence of observation and observational language and provides an overview of the various standpoints with regard to the ontological status of scientific theories. The second part introduces the aspects and interpretations of quantum mechanics relevant to the philosophical analysis in the third part. Three fundamental questions are discussed as part of this analysis: a) what kind of relationship there is between theory and observation in quantum mechanics, b) whether quantum mechanics is an acceptable theory, and c) which ontological status can be assigned to quantum mechanics.

The second part by Annika Schuster is called "Quantum Logic" and is divided into three chapters. The first chapter comprises a review of the fundamentals of classical logic and quantum mechanics necessary for the philosophical analysis. The second chapter discusses arguments for the invalidity of the distributive law in quantum logic based on quantum-mechanical considerations and their defeaters. The third chapter examines Hilary Putnam's approach to quantum logic and the implications regarding meaning and bivalence which he derived from his approach more closely.

Although the two parts of the book were written independently, our conclusions are in a way similar: We both concluded that there seems to be no need to introduce a new ontology or logic in the light of the findings of quantum mechanics. Quantum mechanics seems to be compatible both with a classical ontology and classical logic.

Part 1: Theory and Observation in Quantum Physics

1. Introduction

Quantum physics constitutes one of the major developments in physics during the 20th century and remains current to this day. Ever since the first advancements by the early pioneers of quantum mechanics, it has been the subject of great controversy, often touching on questions of philosophy. Here we will pay particular attention to the relevance of discussions in philosophy regarding the concepts of theory and observation to quantum physics.

The first part will introduce a classification of theoretical and observational terms and sentences following Schurz (2013), and will discuss various problems regarding theory and observation. In particular, we will try to understand the age-old discussion between realism and anti-realism as a question of the reference of theory. We will dismiss several ontological positions and lay out some cornerstones for a new radical empiricist view.

The second part will introduce quantum mechanics for those readers not familiar with it, in simple terms. We will notice some of the differences to classical physics, and discuss some of the empirical phenomena that led to quantum physics as well as the core ideas used to explain them. Further on, we will discuss some of the many interpretations of quantum physics out there and the problem they are trying to solve.

In the final part, we will use the philosophical foundations laid in part 1 to reflect on quantum physics. We will discuss various issues regarding the theoretical structure and observation of quantum physics, as well as asking whether quantum physics is an acceptable viewpoint from a philosophical perspective, and discuss how the different interpretations can be understood ontologically. We will also argue that a sensible ontological position may be relevant for the success of further research in modern physics, in particular the development of a quantum theory of gravity.

Part I. Theory and Observation

This part will lay the philosophical foundations for our discussion of quantum physics. After some opening remarks on the relationship between theory and observation, we will introduce a formal classification of terms and statements. In this context, we will discuss the relevance of ostensive learning to distinguish observational from theoretical language. Next we will discuss various arguments concerning an asserted dependence of observation on theory. The final section is dedicated to the discussion of various ontological positions. We will propose some key aspects on which to build a modern radical empiricism.

2. The Relationship between Theory and Observation

One of the first logical positivists to distinguish clearly between theory and observation was Moritz Schlick, in his General Theory of Knowledge (1918). He argued that scientific theories consist of an observational part and a theoretical part. Theoretical terms gain their meaning from the role they play in theories. The theoretical and observational parts are linked through a correlation between theoretical and observational terms.

The logical positivists had different viewpoints on the exact nature of the separation of theory and observation, some of them changing their minds several times. What many of these viewpoints seem to have in common though, is that observation is considered a certain basis for knowledge, since it is seen as directly accessible through what might be called introspection. Scientific theories could then be used to explain past observations and also predict future ones. This also serves as a method of establishing empirical justification for a theory. An explicit formulation of this would later be known as the deductive-nomological model.

An alternative view point to that of Schlick, was made popular by Rudolf Carnap's Aufbau (Carnap 1979). All theoretical terms are to be constructed from an observational basis. Theoretical terms can be defined through:

(a) observational terms
(b) logical and mathematical terms
(c) already well-defined theoretical terms

Theory can thus be seen as simply observation enriched with mathematics and logic.

Later views started to question whether theory could be traced back to observation through such definitions. The idea of a correlation between theoretical parts and observational parts was reintroduced (cf. Uebel 2011, section 3.4). We will tie in with this idea.

One important consequence of not defining theory through observation, is that the two are logically completely separate. It is impossible to deduce an observational statement from purely theoretical axioms. The same holds true the other way around. A connection between the two is only established by mapping certain theoretical statements with certain observational statements. This mapping can be expressed in the form theoretical sentence "x" corresponds to observational sentence "y". A more complex correlation is, however, also possible. What we mean by this is that it is not necessary to map every theoretical sentence to an observational sentence individually, but rather to establish certain general rules from which such individual mappings can be deduced.

There must also be a mapping between observational sentences and observations, and for realists a mapping between some theoretical statements and states of reality. The problem here is that observations are not sentences and the linguistic representation of sentences is precisely what we are trying to link to it. So, the correspondence between observational statements and observations cannot itself be expressed as a statement in the form *"x" corresponds to "y"*, because y is not a statement. A similar argument is to be made for theoretical statements and states of reality. Any linguistic representation of reality is already a theoretical statement.

That such a correspondence cannot be expressed through language doesn't necessarily mean it can't be learned. In section 3.1.1 we will introduce the concept of ostensive learning. We believe that the correspondence between observational sentences and observation can be learned in such a way. The link between theoretical statements and states of reality will be discussed further in section 5, which is dedicated to the question of ontology.

The role of scientific theories in our view is to yield, through deduction, some theoretical statements which are linked to observational statements through a mapping as we discussed. In physics this may be through prediction of a certain 'state' of a physical system. This state can then be linked to some observational sentence.

Of course usually this link will not be direct. What observation is made depends on how the state is measured or observed. Generally this is done by means of some measuring apparatus. We can, however, describe the state of the apparatus being caused by a certain physical state using theoretical language as well. The apparatus will have some sort of display, whose state will be included in the state of the apparatus, e.g. a computer screen. Naturally the state of the computer screen doesn't tell us anything about whether we are looking at it, in what angle we are doing so, and so forth. This problem could be solved by adding the light emitted by the screen, which reaches our retina into the description. After that, we could also add the state of our brain caused by it, because our conscious response to the light may depend on psychological and neurological factors. In the end, what we'd actually be mapping to observational statements would be physical states of our brain, or possibly other parts of the body if they are deemed to play a role in consciousness.

We are neither saying, that with the current status of neuroscience, such a description is actually possible, nor that scientists should actually proceed in such a way. The presented procedure would, however, yield the most complete explanation possible for any given observation. In practice, many of these steps are left out for reasons of simplicity. There are two reasons this can be done:

(1) We can reasonably assume some device to work in a certain way. For example, physicists usually don't understand or describe the entire measuring device they are using. They simply assume that it displays the results correctly, while always being aware of the possibility that this assumption is wrong. Note that the scientist also doesn't have to understand the exact workings of the measuring device. The only important issue is that he understands which reading on the display corresponds to which state of the physical system being measured. The same goes for our brain. The precise workings of it as well as the precise correspondence of certain states of the brain with observations is still a mystery. Yet, we know roughly what observation to expect when a certain pattern of light enters our eye. But here we also have to be aware of the possibility of our brain not work-

ing precisely the way we expect. The most extreme example would be the possibility of hallucination.

(2) Part of the 'full description' can be left out, because the precise observation is often not of interest. We are rather interested in whether the observation is of a certain type. For example, when reading some measurement on a computer screen, it is irrelevant if the perception of the screen is slightly tilted due to the head of the observer leaning to the left.

An important aspect of our viewpoint is, as with most logical positivists, the notion that our observations can justify observational sentences. They can thus serve as a basis for all science. Karl Popper called this perspective psychologism (Popper 2005). It is opposed to doxasm, which states that all justification of statements must be through deductive reasoning from other statements. Psychologism claims that experiencing something is sufficient justification for the linguistic representation of this experience. Sometimes this is taken to be a *certain* justification. The reasoning for this is that, while you may be experiencing a hallucination and thus not experience reality correctly, you cannot possibly be mistaken about the experience itself.

Popper, in his Logic of Discovery (Popper 2005), further claimed that a theory-independent language of observation is impossible. Instead of observational language, he introduces 'basic statements'. These serve as the most fundamental sentences by which more general theories can be tested. There are two main differences between Popper's view and the one we hold. The first is that Popper doesn't allow for separate theoretical and observational languages. The reason he believes such a separation to be impossible will be discussed in section 4.1. The second regards the justification of these sentences. Popper disregards what he called psychologism. He thinks justification is only possible through logical deduction by other sentences. Instead, he proposes that the 'basic statements' are not supposed to be free from theoretical elements. They must be singular existential statements which can be agreed upon by scientists. They are fallible and may later be rejected. While scientists may be *motivated* by observation to accept certain basic statements, observation does not serve as a justification.

We believe there is a fundamental problem with relying on a consensus for basic statements. The individual scientist has no guideline on the basis of which to decide whether to accept a basic statement or not. Popper agrees that observation serves as a basis for knowledge for the individual, but it may never justify a statement (Popper 2005, p. 74). How can the individual scientist thus know whether to accept a certain basic statement or not? And, if the individual scientists have no criteria by which to decide whether to accept a statement or not, any statement they will be able to collectively agree upon will thus not have been reached by any criteria which might justify giving this statement a somehow heightened epistemic status. All that happens when scientists reach a consensus is that the individuals happen to accept the same sentence, each one of them with no justification whatsoever. We might as well accept basic statements generated by any other means, such as a random generator. Note that basic statements, and thus all of science, don't have any connection to observation at all, other than that they might or might not be a motivating factor for individual scientists. This raises the question, whether science as imagined by Popper can even be considered empirical. Popper's conventionalism may be true in the sense that it describes how scientists actually proceed. If a basic statement is agreed on by everyone, scientists can use it to falsify theories, but this is independent of whether the statement is actually true. In the words of Dennis Graemer:

'Ohne einen gewissen Grad an Einigkeit innerhalb der Forschergemeinschaft kann keine Theorie ihre Geltung beanspruchen. In dieser Hinsicht hat Popper mit seinem Konventionalismus durchaus Recht: Faktisch gesehen werden Basissätze immer durch Beschluss anerkannt. Egal wie falsch, egal wie schlecht begründet ein Satz auch sein mag, wenn er von der gesamten wissenschaftlichen Gemeinschaft als wahr anerkannt wird, so stellt dieser Satz de facto einen Basissatz dar, an dem jene Wissenschaftler ihre Theorien prüfen könnten. In dieser deskriptiven Hinsicht ist die Aussage, dass Basissätze durch Beschluss zustande kämen, geradezu trivial. Aufgabe der Wissenschaftstheorie ist es jedoch, Methoden zu liefern, anhand derer Gründe für oder gegen die Anerkennung bestimmter Sätze abgeleitet werden können.'[1] (Graemer 2012, p. 6)

---

1 'Without a certain degree of unity within the research community, no theory can claim validity. In this sense, Popper's conventionalism is correct: factually basic statements are always acknowledged through a decision. No matter how wrong, no matter how faultily justified a statement may be if it is accepted as true by the entire scientific community this statement will de facto be a basic statement which scientists could use to test their theories. In this descriptive sense the claim that basic statements come about as a decision is almost trivial. But the task of philosophy of science is to deliver methods which can be used to deduce reasons for or against a specific statement.' (translated by the author)

Much more interesting than the notion that statements cannot be justified through observation, seems to be the notion that all statements are necessarily formulated as theory-laden realistic statements. Gerhard Schurz, on whom we will rely heavily for introducing a precise classification of terms and sentences, partly agrees with this. In his conception observational statements are indeed realistic statements, which also have an observational content (Schurz 2013, pp. 107–110). They are, however, not classified as 'theoretical statements'. We believe that his notion of using ostensive learnability to distinguish observational statements, also allows for non-realistic observational sentences, as we will discuss later.

But why do we even need such non-realistic statements? Introducing them into science would be highly complicated and likely have little benefit. We believe them to be beneficial primarily in the realm of philosophy, because it allows for a better understanding of scientific realism and instrumentalism. Realism can then be understood as the notion that the theoretical statements of science refer to actually existing entities, while instrumentalism deems them to be mere tools for the prediction of future observations. A more detailed discussion of the ontological debate between scientific realism and instrumentalism will follow in section 5.

3. Classification of Terms and Sentences

In this section we will introduce the idea of empirical and theoretical terms in the style of Gerhard Schurz (Schurz 2013, pp. 96 ff.), which will later be used to define empirical and theoretical sentences. Empirical and theoretical terms are both descriptive terms. This differentiates them from prescriptive terms referring to questions of moral judgement. They can also be differentiated from purely logical terms, such as logical connectors.

3.1 Empirical Terms

Empirical terms include both observational terms and empirical disposition terms.

3.1.1 *Observational Terms and Ostensive Learnability*

Observational terms refer to some perceivable trait or a complex of such traits. Colours are examples of singular perceivable traits. Observational terms may, however, be much more comprehensive. The term 'raven' for example can be understood as the perceptions associated with ravens, which may include a wide range of properties including shape, colour and sound.

As such, observational terms are terms whose meaning is learnable ostensively. A term is learnable ostensively if it can be learned by being pointed out positive and negative examples of the term, i.e. by inducing the learner with several observations and informing him which of them constitute positive and which negative cases of the term.

The observational usage of the word raven, as in the example above, differs from the way a biologist might use the term. A biologist will likely refer to a raven as a member of a taxon in the sense of some taxonomical theory. The taxonomical theory may refer to some criteria which are not observational. In the cladistic approach for example, vital importance is given to ancestry, which can obviously not be directly observed, but has to be reconstructed using genetic and other evidence.

Schurz (2013) understands observational terms to have a realistic content. But we believe that the concept of ostensive learning also allows for observational terms free of any realistic implication. Since the terms are learned through observations, there is no reason to start associating any more than observational characteristics with them. If someone is taught the term 'raven' he will be confronted with certain observations. There is no reason why he should thus associate anything else but certain observational features (e.g. really existing, mind-independent objects) with 'raven'.

What is, however, required for ostensive learning to work, is an intersubjectiveness of observation. If I wish to introduce a new colour-term for example, I have to assume that other people experience the same objects in a similar manner. Let's say I wish to teach the term "blue" to someone who is not familiar with it. I will point out several objects and comment which ones are blue and which ones aren't. To me the blue objects obviously have some similarity, which is why I recognise them as belonging in the category 'blue objects'. It is theoretically possible that someone else doesn't experience these objects in the same way. He might experience some of them 'red'. In that case he will not be able to learn the new term.

This doesn't mean that different people have to experience 'blue' in the same way. There is no way to test that, because we have no access to other people's minds. For ostensive learning to work, all we need to assume is that we experience certain similarities between the same objects.

While realism might be seen as the best explanation of such an intersubjectiveness of observation, it is not necessarily implied by observational terms. Realism will be discussed in section 5. The possibility of theory-independent observational language will be further discussed in section 4.1.

3.1.2 Empirical Disposition Terms

An empirical disposition term expresses the disposition to react to some condition in a certain empirically observable way. An example is 'water-soluble', assuming that the dissolving of the substance can be observed. 'x is water-soluble' is equivalent to the general law hypothesis: 'Whenever x is put in water, it will dissolve.' All empirical disposition terms can be translated into such hypotheses.

Empirical disposition terms transcend the actual observable since one cannot observe whether the disposition really does hold true in all possible cases. However, they do not reference theoretical entities, but only possible observations. Consequently they should be seen as empirical terms.

3.2 Theoretical Terms

Schurz differentiates between theoretical terms in a broad and narrow sense.

3.2.1 Theoretical Terms in the Broad Sense

Theoretical terms in the broad sense (s.l.) are any descriptive terms which are not empirical terms, but are nonetheless meaningful. They are not limited to terms which are part of a scientific theory but may include metaphysical or speculative terms, if we consider such terms to be nonetheless meaningful.

3.2.2 Theoretical Terms in the Narrow Sense

Theoretical terms in the narrow sense (s.s.) are theoretical terms (s.l.) which are introduced as part of a scientific theory. They thus exclude any metaphysical terms, but are limited to scientific theories with empirical relevance. If a theoretical term (s.s.) is introduced as part of a theory T, it is called T-theoretical. Examples of theoretical terms in physics include force, energy and current. For simplicity from now on 'theoretical term' will always refer to theoretical terms (s.s.), as these will be more relevant for our discussion of quantum mechanics.

3.3 Classification of Statements

This subsection will discuss the classification of statements offered by Schurz (2013, pp. 107–110). We will largely adopt it, with one minor difference, which is due to the fact that Schurz consideres observational statements to have realistic content and we don't. A discussion of this divergence is made in section 5. We will first present the classification offered by Schurz, then explain his reasons for adopting it and why we need to make a slight adaptation. As we did with terms we will limit ourselves to the synthetic-descriptive realm.

The classification of the sentences is based on the type of terms they use, as well as their logical form. Schurz's classification goes as follows:

'If S is a synthetic-descriptive sentence, the following holds:

> a. S is an *observation sentence* iff apart from logical concepts, S contains only observation concepts, and S is either a singular sentence, or a *localized* existential or universal sentence, whose quantifiers possess empirical range.
> b. S is an *empirical sentence* (i.n.s.) iff apart from logical-mathematical concepts, S contains only empirical concepts (i.n.s.), and its quantifiers possess empirical range.
> c. S is a *theoretical sentence* iff S contains theoretical concepts i.w.s. *or* its quantifiers possess theoretical range.
> d. S is a *T-theoretical* sentence of (theoretical i.n.s.) iff S is a theoretical sentence i.w.s. and the theoretical concepts of S are introduced by the scientific theory T'
> (Schurz 2013, p. 108)

The existential or universal statements that make up observational statements must be localised. This allows for an observational examination of their truth. So an existential statement would have to specify the location of the object, while a universal statement would have to be limited to a region which can be observed entirely.

Empirical reach means that only observable objects are included. The reason observational statements have to be limited to those objects that can be observed is obvious. The statement 'All objects in this room are red.' cannot truly be considered observational if it includes molecules so small that they cannot be observed, but which nevertheless reflect light with a wavelength corresponding to the colour red.

For us, the redness which an unobservable object might have, i.e. the disposition to reflect light with a certain wavelength, is, however, not the same as the observational term 'red'. The observational term simply refers to red observations. Schurz on the other hand takes observational terms to have a realistic content, which is why the observational term 'red' may include dispositions which also apply to non observable objects. So, for Schurz the limitation 'empirical reach' may be necessary, but for us it is irrelevant. Unobservable objects cannot be red in our sense.

4. Theory (In-)dependence

A number of arguments have been put forward to establish a dependence of observation on theoretical concepts (cf. Schurz 2013, 65–72). We will distinguish between two sorts of arguments and discuss their accuracy and relevance to the view on theory and observation mentioned above. The first sort of arguments state that all descriptions of observations do include references to theoretical concepts, thus putting into question any clear distinction between observational and theoretical statements, as the one we adopted in the previous section. The second sort refers to a supposed dependence of the observations made on theoretical assumptions. This includes the notion of linguistic relativism, which we will, however, discuss lengthier in a separate subsection.

4.1 Theory (In-)dependence of Observational Language

The point that it is impossible to formulate a theory-independent sentence was most notably put forward by Karl Popper in the Logic of Discovery:

> 'Wir können keinen wissenschaftlichen Satz aussprechen, der nicht über das, was wir >auf Grund unmittelbarer Erlebnisse< sicher wissen können, weit hinausgeht (>Transzendenz der Darstellung<); jede Darstellung verwendet allgemeine Zeichen, Universalien, jeder Satz hat den Charakter einer Theorie, einer Hypothese. Der Satz: 'Hier steht ein Glas Wasser' kann durch keine Erlebnisse verifiziert werden, weil die auftretenden Universalien nicht bestimmten Erlebnissen zugeordnet

werden können (die >unmittelbaren Erlebnisse< sind nur *einmal* >unmittelbar gegeben<, sie sind einmalig).'² (Karl Popper 2005, p. 71)

'Here is a glass of water' is, according to Popper, a realistic statement. It transcends the mere observation and references the real existence of a glass. Even if we were to replace the sentence with 'I have the perception of a glass of water', it still contains reference to a glass, which is a theoretical and realistic concept.

Schurz's definition of observational terms as those that can be learned ostensively offers a way out of this problem. Schurz himself does take the observational statements to have both a realist and an observational content (Schurz 2013, p. 65), but ostensive learning relies solely on observations. What is learned is thus only observational, too. When the learner is induced with certain perceptions, e.g. of a raven, there is no reason he should identify anything else with the raven than certain observational characteristics. An interpretation of observational statements as free of realistic content is thus also possible.

We should not dismiss Poppers objection to the notion of pure observational language completely though. While ostensive learning doesn't require the learner to associate any realistic content with the terms he is learning, he does need to make certain theoretical assumptions. If we show him a black raven flying in the blue sky, we assume that he will recognise the black part of his visual perception as a distinct object from the blue background. Such a recognition of individual objects already requires some theoretical background. We concede that this is the case. However, the individual object, or 'thing', doesn't have to be interpreted in the realist sense. In the 'realistic interpretation, this term refers to assumed entities outside of one's experiences, and in the phenomenalistic interpretation it refers to entities within one's experience.' (Schurz 2013, p. 65)

---

2 'We can't speak any scientific statement which doesn't go far beyond what we can know for certain >on the basis of direct experience<. (>transcendence of representation<); every representation uses general symbols, universals, every statement has the character of a theory, an hypothesis. The statement: 'Here is a glass of water' can't be verified by any experience because the occurring universals can't be matched with any specific experiences (the >direct experiences< are only once >directly given<, they are unique).' (translation by the author)

Let us be a bit clearer on the role of theoretical assumptions in ostensive learning. The learner is confronted with several positive and negative cases of the term to be learned. How does he, from some particular examples, learn an abstract term? The goal is, that he be able to recognise actualisations of the term, even those he has not been confronted with before. After the learning process he should be able to tell which observations constitute an example of the term, even if he has not previously been confronted with an identical observation. In order to do this he has to form theories about what is meant by the term. These theories can then be empirically tested through the positive and negative examples he is given.

What kind of theories he will come up with, will depend heavily on his previous beliefs, like what kinds of features of the observations he deems relevant or what sort of observations he considers to be similar. In fact similarity seems to be very crucial, since an important aspect in ostensive learning is recognizing the similarities between the various positive examples.

What is not important in ostensive learning, is the belief that the observations are caused by 'real' objects. It can further be argued that, while previous (theoretical) beliefs are important for ostensive learning, this doesn't imply that the learned term itself has theoretical content. Either way ostensively learned terms, do not have to imply reference to 'real' mind-independent entities.

4.2 Theory (In-)dependence of Observation

The second sort of argument claims that what observations are made somehow depends on theoretical assumptions. Here are some ways in which this is supposedly the case:

(1) The theories we hold determine where we direct our attention and what experiments we conduct.
(2) Perception sometimes depends on pre-expectation.
(3) Our 3D visual observations are constructed by our brain from two 2D images received by our retinas.
(4) Scientific measurement assumes theories about the way the measuring apparatus works.
(5) Observation is language dependent. We can only perceive what is describable in a language we are familiar with.

All of these, except (5) we believe to be true. However, we will doubt that they pose any serious problem for us. Afterwards, we will discuss why we have doubts (5) is true if seen as a separate point from (1).

(1) claims that which experiments we do depends on the theories we accept. This is true, because different theories might raise different questions, so the experiments we believe to be relevant are guided by them. However, it doesn't follow that two scientists who accept different theories will obtain different results if they do happen to perform the same experiment. They will either explain them differently or they may not be able to explain them within their theory, meaning they have to either adapt their theory or abandon it.

Points (2) – (4) are not problematic, because they all refer to some theories, which we need to take into account to correctly predict our observations. This does not contradict the notion that theories are used to explain and predict observations. On the contrary, all of these actually specify theories we need to take into account when making predictions. We will, in order, explain how this is the case for each one.

While perception may sometimes be dependent on pre-expectations, there isn't an absolute correlation. We don't always experience exactly what we expect. I think it is clear how this would lead to an absolute circularity of science. Luckily it is not the case.

And as long as it is not the case, we can view (2) simply as another theory, which needs to be taken into account when making predictions. To be precise, it means that we always have to consider, that our perceptions may be closer to our expectation, than they would be, if (2) was wrong. Scientists can make provisions in order to make sure their observations are not simply the result of their pre-expectation, for example by letting computers evaluate the measurements. As long as their observation of the computer screen isn't also completely deluded by their pre-expectations in a way that they might even see a 1, when there is really a 0, the non-circularity of science is not in danger. And even if this is individually the case, other scientists will point out that they are getting different results.

(3) contradicts naïve realism, the notion that we perceive objects as they really are. We can, however, simply take into account that what we perceive is a construction of two 2D images. If a theory predicts that our retinas should receive certain 2D images, by considering how 3D images

are constructed from the 2D images, we can make predictions about our actual observation.

Unlike (2) and (3), (4) doesn't make us aware of a theory that we should take into account, but rather reminds us of theories we already assume. When performing an experiment to test some new hypothesis, scientists always use equipment in the measurement process, which they assume to work in a certain way. When for example taking a measurement of temperature with a mercury thermometer, what one directly observes is only the expansion or compression of the mercury. We assume that this is related to temperature in a certain way.

So what scientists often call 'data' is in fact not observation, but rather an interpretation of observations based on assumptions about the experimental set up. (4) thus doesn't concern the distinction between theory and observation.

4.3 Linguistic Relativism

The idea of linguistic relativism was made popular by the linguists Edward Sapir and Benjamin Lee Whorf. The core idea is that languages differ in important ways which influence the way we experience and conceptualize the world (cf. Chris Swoyer 2003). While it may well be true that language has an influence on the way we think about the world, we have doubts that it influences our observations, except in ways that can be reduced to (1), which we have shown to be unproblematic. We will build up our discussion of language relativism along the example of the lack of separate words for yellow and orange in the Zuni language (Lenneberg and Roberts 1953).

In what way might language influence observations in the sense of (1)? Zuni native speakers may, when being confronted with an orange object, not have a closer look to see what precise colour it is. Since yellow and orange are expressed by the same word in the Zuni language, it may not be relevant for them to find out which one it is. This corresponds to (1), since it is the theoretical assumptions about the categorization of colour that leads the Zuni speaker to not direct his closer attention to the colour of the object.

Whether language influences the observations we make if we do direct our attention to a certain object is an empirical matter. We wish to demonstrate what dramatic empirical consequences it would have if Zuni speakers really didn't see a difference between orange and yellow. If we were to generalize this principle it would also imply that english-speakers who know the word "blue", but not words for different shades of blue, experience all blue colours the same. It could never be the case that they experience two different shades of blue, without knowing what they are called. For example if you don't know the words "Royal blue" and "medium blue", you'd experience them identically.

The only difference between Zuni-speakers and English-speakers who don't know words for different shades of blue is that in English these different words do exist. Some English-speakers are simply not aware of them. So unless the existence of words for colours in the language you speak, but which you are not aware of, can influence the way you experience a colour we should really expect these consequences. This is, however, highly implausible, and we think many people will testify that these consequences contradict their personal experience.

We don't have to rely on such personal testimony, though. Schurz (2013, p. 71) made the point that if people who are not aware of two different words for orange and yellow were to perceive them the same, they shouldn't be able to learn such words ostensively.[3] If they were shown yellow and orange objects and then told which ones are yellow and which ones are orange, they couldn't tell the difference, so would thus later not know when to use the word for orange and when the word for yellow. This is, however, not the case. Speakers of the Tiv and Dani languages were able to learn colour concepts for which there is no word in their native tongue ostensively (Garnham and Oakhill 1994, pp. 39–51; Berlin and Kay 1999, p. 24).

A weaker version of linguistic relativism, in which the difference between colours for which one does not know a name is experienced, but less intensive, may still be true. As long as there is some small difference one would in principle be able to learn words for the different colours ostensively. The burden of proof for this rests with the proponents of such a theory though and if it were true, it would not pose a serious problem

---

3   Compare section 3.1.1 for an explanation of the concept of ostensive learning.

to the distinction between theory and observation. While this would be another example of the influence of ones theories on ones observations, comparable to (2), it does not imply that theory and observation aren't in principle distinguishable concepts or that science is a circular enterprise.

5. Ontological Status of Scientific Theories

There is a variety of positions on the ontological status of scientific theories. The most important divide in ontology is been between realism and idealism (antirealism). Historically ontology has often been concerned with the status of the objects of our everyday experiences. In philosophy of science, the focus is on the status of scientific theories and the theoretical entities they entail. Here the main positions have been named *scientific realism* and *scientific instrumentalism*. In the first subsection we will attempt to define (anti-)realism and distinguish the debate from other philosophical issues. The second subsection introduces different kinds of antirealism and realism. Next we will argue against various of these positions. Finally we will discuss our own position, which is in the tradition of the logical positivists.

5.1 Realism and Antirealism

Realists believe that there is some sort of reality which is independent of the mind. Objects that are part of this reality exist independently of our observation of them or our theories about them. Antirealists either deny this or claim that the notion of mind-independent existence is ill-defined. They believe that only the content of minds can be said to exist, i.e. conscious experiences, observations, thoughts, beliefs and ideas. A more differentiating account of the sort of things antirealists believe to exist will be given in the next section.

Epistemological realism is logically stronger than this (ontological) realism. In addition to claiming that a mind-independent reality exists, it is claimed that we can gain knowledge about this reality. While epistemological realism can be seen as a position on epistemology rather than ontology, it implies ontological realism. It makes no sense to claim we have access to mind-independent reality without also believing that such a reality exists. But it remains important to note that whether something exists and what we can know about it are two different issues. One may well be an ontological realist without being an epistemological realist. In that case one believes that there exists something which is independent of the mind, but we can never know anything about it.

Scientific realism and instrumentalism deal specifically with the ontological status of scientific theories. The former holds that the theoretical entities which are part of our scientific theories refer to some real entities. The exact nature of the relationship between the theory and these entities is subject of debate.

Recall the classification of terms presented in section 3. Observational terms are those that can be learned ostensively. Unlike Schurz, we consider them to have purely observational content, i.e. the observable traits associated with it. They may include theoretical interpretation, but only of observations, not mind-independent objects. Distinct from the observational terms are the theoretical terms, which don't simply reference observations.

With such a classification in mind we get a clearer notion of scientific realism and what its rejection implies. Rejection usually implies the belief that the content of science can be reduced to (potential) observations. Only observational statements refer to real entities, i.e. observations.

Instrumentalists believe that the goal of science is simply to explain and predict observations. Theories are selected for their utility for making correct predictions. Their theoretical entities are not seen as referring to real entities, or at least this is not seen as relevant.

One might think that realism and scientific realism have to go hand in hand and that the same is true for antirealism and instrumentalism. But this is not necessarily the case. Scientific realism entails realism, because it postulates the existence of that which scientific theories are thought to refer to. However, an ontological realist may believe that a mind-independent reality exists, but that scientific theories don't refer to it. This might be the position of an ontological realist who is not also an epistemological realist.

For the same reason, instrumentalism doesn't necessarily imply idealism. One may, as a realist, believe that scientific theories don't refer to reality, but that we can still use it to predict observations. On the other hand idealism does seem to imply instrumentalism, unless someone suggests an option that is distinct from both scientific realism and instrumentalism, since idealism is certainly not compatible with scientific realism.

These considerations demonstrate the difference between the ontological question 'Is there a mind-independent reality' on the one side and the epistemological question 'What can we know about mind-independent reality?' on the other side. In philosophy of science the epistemological question addresses scientific theories in particular. What do they tell us about a mind-independent reality? All of these questions are interrelated though, and will be considered in this section.

Another issue which needs to be distinguished from ontology is that of the correspondence theory of truth, which states that the truth of a statement is determined by a correspondence of the statement with reality. The statement 'There exists a black raven.' is true if and only if there really does exist a black raven. One can thus argue, that for any statement to be true there has to be a reality with which the statement can correspond. This is true; however, it tells us nothing about the content of this reality. Is it limited to the content of minds or does it include mind-independent objects? An idealist can thus also accept the correspondence theory of truth, only that he believes the reality with which a statement might correspond to be limited to minds. The correspondence theory of truth is thus 'metaphysically neutral' (cf. Kirkham 1992, sec 6.5). It does, however, raise the question whether an antirealist can consider theoretical statements to be true, since antirealists don't believe them to refer to anything with which they might correspond. Instead of regarding theoretical statements as truth carriers, they could be described as more or less 'useful' in predicting observations. The correspondence theory of truth can, however, still be applied to observational statements.

5.2 Kinds of (Anti-)Realism

This subsection will introduce some important ontological positions. The first part will introduce more fundamental positions and more complex views, which are often based on these fundamental positions, will be discussed in the second part.

5.2.1 Fundamental Positions

There are three important kinds of idealism (Dancy 1985, p. 136 ff., Schurz 2013, p. 60):

Subjective idealism/Solipsism: Only ones own experiences exist.
Intersubjective idealism: Only ones own experiences and those of other subjects exist.
Possibilistic idealism: The possible experiences of all subjects exist.

While all idealists agree that only experiences exist, they don't agree on what subjects there are that might have experiences and on whether possible experience can be said to exist, too. Solipsism can be seen as the most restrictive form of idealism. A solipsists believes only in his own experiences. Intersubjective idealists acknowledge that there are other subjects than themselves. Opinions on what other subjects exist vary, e.g. on whether non-human animals may be subjects. Note that the recognition of other subjects doesn't imply that one has access to the experiences of other subjects. Precisely this makes it hard to recognise other subjects as subjects, though. Possibilistic idealists also believe possible experiences to exist in some sense. At least they believe statements about such experiences to be meaningful. If there is a table inside a room with no person inside, a possibilistic idealist might say, 'If someone was to enter the room, they'd see a table.', even if no one actually does so.

Epistemic Realism can be divided into two forms (Schurz 1995):

Naive Realism: Reality is exactly as we experience it.
Indirect Realism: Reality is not exactly as we experience it, but we can indirectly gain knowledge about it through our experiences.

Naive realism is largely disregarded now, which is partly due to the fact that our 3D visual observations are constructed by our brain from two 2D images received by our retinas, a fact we already mentioned in section 4.2. However, most realists believe that we can gain knowledge about reality through our experiences, even though our experiences cannot be said to be an exact, direct representation of reality.

Other fundamental positions which we already discussed in the previous subsection are ontological realism, scientific realism and instrumentalism. It should be noted that it is also possible to take a scepticist stance on most of these issues. One might for example be a scepticist towards the question of the existence of other subjects. In that case one doesn't deny such an existence, as a solipsist does, but one believes that we cannot know whether other subjects exist or not. The same can be said for realism. A scepticist may not only argue that we have no knowledge about the truth of certain realistic statements, such as realistic interpretations of scientific theories, but also that we can have no knowledge of whether such a reality exists. Such a position is also opposed to idealism which states that such a reality doesn't exist or that the concept is ill-defined.

5.2.2 Advanced Positions

Two specific antirealist positions we would like to discuss are *radical constructivism* and *logical positivism*. While both are antirealist, there are significant differences in their argumentation against realism. We will discuss them in turn.

Radical constructivists argue that our perceptions are not of a reality as it is. Instead, they are constructions by our mind or brain. They conclude from this that there can be no discoverable reality independent of our mind and further no mind-independent reality at all. One could say that a rejection of naive realism leads them to rejecting epistemic realism altogether, which in turn leads to a rejection of ontological realism or at least scientific realism. We will examine the validity of this argumentation in section 5.5.

Logical positivism is an umbrella term for a range of views on the philosophy of science, most notably those held by members of the *Vienna Circle*. While their views on ontology varied, a common theme seems to be the notion that *real existence* of mind-independent objects is a meaningless concept, since it doesn't have empirical content. Not only does it not add predictive power to a theory, but it is not even understandable, since it cannot be reduced to empirical terms. Consequently logical positivists don't claim that there is no mind-independent reality, but rather that the claim of its existence is meaningless. The negation of it is thus just as meaningless. We will argue for a modern version of logical positivism in section 5.7.

Next we will discuss two opposing realist positions. An important distinction is to be made between *metaphysical* realism and *hypothetical* realism (Schurz 2013, p. 293). Both are forms of scientific realism. Let us explain what we mean by metaphysical and hypothetical realism, before we will introduce *hypothetical-constructive* realism, as an example of hypothetical realism.

For a metaphysical realist, scientific realism, i.e. the notion that theoretical entities refer to real entities, is 'a *necessary* precondition of theoretical science' (described, but not endorsed by Schurz 2013, p. 293), a view we will argue against in section 5.4. Hypothetical realists view scientific realism as a 'fallible hypothesis' (Schurz 2013, p. 63), which relies on justification through its *'empirical success'* (Schurz 2013, p. 61).

Hypothetical-constructive realism holds that the tie between a true statement and the object it refers to is not a 'quasi-identical mirror-image', but a 'structural correspondence' (Schurz 2013, p. 62). The relationship is thus neither exact nor unique. A true statement is merely 'an incomplete structural representation' (Ibid.) of some aspect of reality. Schurz puts emphasis on constructive realism being a hypothetical realism (Schurz 2013, p.293 ff.). The hypothesis of realism is not given a priori. but requires justification through abduction, i.e. the inference of the best explanation. The explanandum is the empirical success of scientific theories. We will discuss the justification of constructive realism further in section 5.6.

5.3 The Case against Solipsism

Solipsism often relies on sceptic arguments that question our ability to have knowledge of other minds. We may be able to observe the physical body of other people, the way they behave and communicate with us. But none of this ever gives us access to their minds. So how do we even know they have one? Such sceptic arguments cannot establish a strong version of solipsism that claims we can have knowledge that there are no other minds. Just because we don't have access to other minds, doesn't mean they don't exist. A sceptical version of solipsism, that simply states that we cannot know whether other minds exist or not, is, however, much more difficult to refute and arguably there will always remain some justified scepticism.

The most promising approach to establishing other minds seems to be that of analogical inference (cf. Alec Hylsop 2014, sec. 3.1). The idea is that we experience other people to be like us in various ways, including their physical characteristics, behaviour and the thoughts they communicate. Since these seem to correlate with certain states of mind for us, it seems plausible that the same holds for others. Modern proponents of this argument often bring it in the form of a scientific inference. Admittedly, the empirical basis for such an inference is relatively small compared to what is trying to be established. We only have access to one person's mind, our own, yet we intend to establish the existence of minds of billions of people and possibly even non-humans. It does, however, yield us a certain level of justification.

5.4 The Case against Metaphysical Realism

As discussed in section 5.2.2 metaphysical realism holds that realism is a necessary presupposition of science. One way of arguing this is by building the discovery of mind-independent reality into the goal of science. This seems to be a cheap trick because with it we can establish almost anything, simply by claiming it is presupposed by the goal of science. What many should be able to agree on though is that part of the goal of science, if not the goal of science, is 'to find *true* and *content-rich* statements, laws, or theories, relating to a given domain of phenomena' (Schurz 2013, p. 19, also cf. Weingartner 1978, sec. 3.2.). But that truth, particularly the correspondence theory of truth, does not require correspondence with a *mind-independent* reality was already established in section 5.1.

The existence of antirealist approaches to science puts into question the necessity of realism for science on a more general level. If science really does presuppose realism, one should be able to point out exactly why these antirealist approaches are unsatisfying. While one might disagree with them on various points, including the justification of their antirealism, the implication that science is impossible on their premises seems to be much stronger and requires additional justification.

A further argument against (metaphysical) realism, named *Pessimistic Meta-Induction* draws on empirical evidence from the history of science (cf. Laudan 1996). Many scientific theories that have been empirically successful are now considered outdated and it is no more believed that their theoretical entities have counterparts in the real world. Yet they were empirically successful nonetheless. So the finding of empirical successful theories doesn't seem to require correspondence with mind-independent reality. An example for an empirically successful theory which was later replaced, given by Laudan, is the phlogiston theory, which is now dismissed in favour of the oxygen theory.[4] We will return to the Pessimistic Meta-Induction argument in our discussion of constructive realism in section 5.6.

5.5 The Case against Radical Constructivism

As presented in section 5.2.2, radical constructivism concludes from a rejection of naive realism that reality itself is constructed. This seems to be quite a leap. Schurz (2013), argues that the 'subtle Error' is the failure to recognise that the concept of 'knowing' doesn't have to be understood 'in the *naive-realistic* sense of a kind of direct mirror-imaging' (Schurz 2013, p. 62). Instead we should understand 'the relationship between a true statement and its object not as a quasi-identical mirror-image, but as a *structural correspondence*, which *conveys certain information* and neither has to be complete or unique' (Ibid.).

The case of the constructive realists like Schurz, drawing partly on arguments put forward by other scientific realists, is a strong one. It will be the subject of the next section. We will suggest that arguments in the framework of radical constructivism are not sufficient to convincingly reject constructive realism.

5.6 The Case for and against Constructive Realism

5.6.1 *Inference of the Best Explanation*

For constructive realists, a structural correspondence between an empirically successful scientific theory and reality is justified as the best explanation of its empirical success. A well-known version of this argument, the *No-Miracles Argument*, was put forward by Putnam (1975, p. 73). According to this argument, the empirical success of science would be an incredible miracle if not for scientific realism. It is thus more plausible to accept some form of scientific realism. The problem with *inference of the best explanation* or *abduction* is that they are not valid in the strictly logical sense. This raises the question of its justification, which will be discussed in section 5.6.4.

---

4 The original oxygen theory has today also been replaced by a more generalized theory of oxidization.

5.6.2 Pessimistic Meta-Induction

Pessimistic Meta-Induction, which was already mentioned in section 5.4, can be seen as a counter-argument to the No-Miracles Argument. If empirical success justifies the belief in a (structural) correspondence with reality, how come originally empirically successful theories had to be revised so often in the history of science? *Meta-Induction*, i.e. the use of an induction for evaluating methods, suggests that, since the No-Miracles Argument or inference of the best explanation are not very successful, we are not justified in using them. It is important to stress that this is an empirical argument based on evidence from the history of science. It thus depends on the empirical claims it is based on, i.e. that historically a significant number of empirically successful scientific theories have eventually turned out to be wrong.

5.6.3 Intertheoretical Correspondence

Structural correspondence offers a way to deal with Pessimistic Meta-Induction. Even if theories are using 'false ontologies', i.e. they postulate theoretical entities which don't correspond to real objects, there may still be '*something true* in their theoretical superstructure' (Schurz 2013, p. 296). Constructive realism doesn't treat true statements as a mirror image of reality, but as having a structural correspondence with it. Therefore there may be some structural correspondence between a theory and reality, even if the theory is false and another theory shows a stronger correspondence.

The idea is that if a new theory replaces an old theory and entails all of the old theory's empirical success, then the correspondence with reality of the old theory may be conserved, even if the ontologies of the two theories contradict each other. This in turn leads to a correspondence between the old theory and the new theory. Correspondences between successive scientific theories have been pointed out by Boyd (1984) and Worrall (1989/97).

Laudan's responded by arguing that the cases given are mere exceptions and offers a list of counter-examples (Laudan 1996, p. 121). In these examples, he argues, the successive theories have incompatible ontologies and it is therefore impossible that there is any correspondence between their theoretical structures. What he seems to fail to realise is that while two different theories will almost always have different, incompatible ontologies, this doesn't necessarily mean that parts of them can't show some similarity in the structure. This similarity can be understood in terms of a correspondence relations.

Schurz (2009) goes so far as to claim the opposite of Laudan. He offers logical proof for his *Correspondence Theorem* based on assumptions he deems plausible. The Correspondence Theorem states;

> 'if a physical system x satisfies the conditions A (which define the range of [the theory T] and [its successor theory] T*'s shared empirical success), then x satisfies the T-theoretical description τ iff x satisfies the T*-theoretical description τ^*.' (Schurz 2013, p. 297)

This means that some structural correspondence between two successive theories which share some empirically successful predictions is necessary. Schurz, goes on to show how one of the conflicting ontologies examples given by Laudan, the succession of phlogiston theory by oxidation theory, does in fact include a correspondence relation.

There are two reasons Schurz's Correspondence Theorem is relevant. First of all, it follows that the Pessimistic Meta-Induction argument fails, because in all cases of succession of theories some structure of the original theory is conserved. The replacement by a successor theory thus doesn't necessarily mean that the original theory didn't correspond to reality in some way. It might be that the structure that has been conserved is actually one corresponding to some aspect of reality. The second reason Schurz's Correspondence Theorem is relevant is that it serves as a basis for his attempt at justifying abduction or inference of the best explanation, as we will see in the next subsection.

5.6.4 Justification of Abduction according to Schurz

Schurz (2009, sec. 7.3), continues by arguing that the correspondence theorem may provide a justification of the abductive argument for realism, if a further assumption is made. The correspondence theorem itself only implies that if the present theory is true, there is some truth to a preceding theory which shares some of the empirical success of the current theory. But instead of considering a current theory, we can apply it to an *unknown ideal theory*'(Schurz 2013, p. 299). Schurz calls the assumption that such an ideal theory, which offers 'an approximately true description of the structure of reality' exists, *minimal realism*. If we apply the correspondence theorem to this ideal theory, it follows that strong empirical success of any theory implies partial realistic truth.

Since Schurz has to make the assumption of minimal realism, his argument can be attacked by rejecting his assumption from the start. Nonetheless, his argument serves as a powerful counter to the radical constructivists, since their arguments are critical of our ability to gain knowledge of reality. Schurz managed to demonstrate that, assuming an ideal theory describing the structure of reality exists, we are justified in assuming our empirically successful theories entail some truth. So the assumption that we can have no access to reality, made by the radical constructivists, is put into question.

5.6.5 The Logical Positivist Objection

A typical logical positivist position is to deny that the notion of 'real existence' of theoretical entities is meaningful. Meaning has to be related to observation. However, the notion of 'real existence' doesn't add empirical content to a theory. This objection can be raised to constructive realism, too. Here the talk is of 'structural correspondence' of theories with reality. Nonetheless this correspondence doesn't add empirical content, if this reality is understood as mind-independent.

A logical positivist thus also has to reject Schurz's justification of abduction, since it is based on the assumption of minimal realism. Minimal realism presupposes the existence of a reality, which is described by an ideal theory. But it is the meaningfulness of the notion of such a reality which is already being denied.

It might be argued that, even if we cannot rely on Schurz's justification of abduction, scientific realism can be established through scientific inference. If this inference should not be justified, this is not just a problem of realism, but of science in general. This argument is independent of what precise form we believe such an inference to have, e.g. abduction, induction or critical testing as suggested by Popper (2005).

The difference between the use of such arguments to justify realism, and the use of scientific inference in empirical science, is that a realistic theory doesn't have more empirical content than its non-realistic counter-part. In science an empirically successful theory may be explained by a more general theory. But this general theory has to explain additional evidence or make additional predictions, not already given by the former theory. With realism this is not the case.

Realism is in this regard no different than the metaphysical explanations criticised by the Vienna Circle. Among these were vitalist theories, which attempted to explain aspects of living organisms by a *vital force* or *entelechie*, as Hans Driesch called it (cf. Verein Ernst Mach 1981, p. 312). The idea is that living organisms have certain characteristics such as growth or the ability to move because of a vital force. Sometimes vitalists were able to give accurate descriptions of empirical correlations. These were to be explained by vitalism.

The logical positivists argued that terms such as *vital force* cannot be constituted from the observable and are thus meaningless. We can disregard vitalism for a similar reason, without having to accept a verificationist criterion of meaning. The hypothesis of vitalism simply doesn't add any empirical content to the mere description of observable correlations, on which vitalists and non-vitalists often agreed. Vitalism tells us nothing about the way an organism moves or grows, beyond what we already know and are trying to explain. It should be easy to accept that such an explanation is useless at best.

The same can be said for realism. When assuming 'real existence' of the theoretical entities of some theory, we add no empirical content to our theory at all. There are no predictions we can make with the assumption of 'real existence' that we can't also make without it. We should elaborate a bit further, since we have to recognise that for proponents of the No-Miracles argument, scientific realism doesn't attempt to explain any particular empirical regularities, but rather that there are such regularities at all. Without such regularities science could not be so successful. But the response to this is in principle the same: What does scientific realism tell us about these regularities except that they exist? And that they exist we already know. We have to acknowledge that there are some things for which there is no scientific explanation. They are brute facts, which we simply have to accept based on the empirical evidence. If we don't accept this, we will soon be postulating all kinds of metaphysical entities and additional metaphysical entities to explain these and so on. A similar view is expressed by Van Fraasen in *The Scientific Image* (1980, p. 19 ff.).

5.7 Towards a Modern Radical Empiricism

Based on our considerations so far, we believe that, while we may need to reject some typical logical positivist views, the case against realism is still a very strong one. We should thus work towards a modern version of radical empiricism, drawing on the ideas considered here, including the concept of ostensive learnability as a basis for observational language. An important work on radical empiricism, which should also be taken into account, is *The Scientific Image* by Van Fraasen (Ibid.). It is beyond the scope of this article to present such a modern radical empiricism in detail. Van Fraasen offers a much more comprehensive account. We will simply present four points which we think can serve as a fundament of a radical empiricist outlook. These will be sufficient for our discussion of quantum mechanics. Readers familiar with Van Fraasen will notice some fundamental deviation from his work. A detailed account of our differences with Van Fraasen may be the content of future work.

5.7.1 The Independence of Theory

Unlike some other radical empiricists, we take theory to be relatively independent from observation. Theoretical terms don't have to be constituted through empirical terms. Instead, certain theoretical states are correlated with certain observations. The designation of these correlations is not intended to be a definition. It only links the theory to observations, so that it can actually make empirical predictions, rather than simply predictions about purely theoretical states. We discussed this link in more detail in section 2.

5.7.2 Observational Language as Ostensively Learnable Language

Following Schurz (2013), we introduced observational language as ostensively learnable language in section 3.1.1. We believe that ostensive learnability serves as a good criterion, because ostensive learning depends on observable differences. We can thus use it to really capture observational concepts as opposed to theoretical concepts.

5.7.3 The Limited Theory-Ladeness of Observational Language

As we discussed in section 4.1, observational language does contain some theoretical content, even when defined through ostensive learning. But the important thing for us is solely that it doesn't have to contain *realistic* content. While we might make certain theoretical assumptions when interpreting our observation, we don't have to assume realism. Even if there are some problems separating observational from theoretical content, there can be little doubt that these are in principle different.

5.7.4 Literalness of Theories

One Point which we haven't discussed so far is the *literalness* of theory, suggested by Van Fraasen (1980, pp. 9–13). Van Fraasen makes a distinction between two types of antirealism:

> 'The first sort holds that science is or aims to be true, properly (but not literally) construed. The second holds that the language of science should be literally construed, but its theories need not be true to be good. The anti-realism I shall advocate belongs to the second sort.' (Ibid., p. 10)

'Orthodox' instrumentalists believe that there is no relevant difference between two theories with identical empirical predictions. They 'hold that two theories may in fact *say the same thing* although in form they contradict each other' (Ibid, pp. 10–11). They share the same truth-value, because of their identical empirical content. By taking a theory literally, one rejects this view and accepts it as opposed to any theory with a distinct theoretical structure.

Unlike realists Van Fraasen, however, doesn't take theories to be literally *true*. They don't refer to *real* entities. The theoretical structure is nonetheless deemed important, because accepting a particular structure corresponds to subscribing to a particular research program. Research based on one theoretical structure will lead in a different direction than research based on another structure. So even if two theories make identical predictions at the moment, further research may develop them in different directions.

One such example where the choice of theoretical structure matters is the combination of different theories. When combining two theories we equate some of their theoretical entities. The way we can combine them, and this may affect the empirical adequacy of the resulting theory, may well depend on which theoretical entities and structure we choose for the theories. In section 13.3 we will suggest that this may be relevant for the formulation of a theory of quantum gravity.

We think that Van Fraasens' viewpoint offers a sensible middleground between 'orthodox' instrumentalism and scientific realism. It does not include *reference to an independent reality*, which we deemed to add no empirical content in section 5.6.5. At the same time it allows us to recognise theories with distinct theoretical structures as significantly distinct from one another.

Note that although we believe theories should be taken literally, this doesn't mean that in principle the goal of science is something other than to find empirically adequate theories, as suggested by the scientific instrumentalists. This view is thus not entirely contradictory to instrumentalism. The suggestion is simply to take theories literally in order to be better able to generate more complete and empirically adequate theories.

Part II. Basics of Quantum Physics

This chapter will attempt to give a brief introduction to quantum physics. It is beyond the scope of this article to give a detailed explanation of the mathematical structure of quantum theory, so instead we will simply give an incomplete overview of the most important aspects. Hopefully this will be comprehensible to laymen and equip them with the basic knowledge necessary for understanding our discussion of the concepts of theory and observation in quantum physics, which we will present in the next chapter.

We will begin by considering the picture of the world presented by classical physics. Afterwards we will consider some of the experimental results that started to shatter this picture and the explanation offered by quantum mechanics. We will introduce the restrictions on interpretations of quantum physics imposed by Bell's Theorem and the Kochen-Specker Theorem. Finally, we will discuss some of the many alternative interpretations.

6. The Classical Picture of the World

In classical physics, a system is always thought to have a certain state and several properties. These properties are independent of measurement. While a measurement constitutes an interaction with the system and may thus influence the system in a way which is also described by classical physics, the properties are nonetheless thought to exist prior to and independent of the measurement.

There may be several mathematical formalisms describing the same issue. Lagrangian mechanics and Hamiltonian mechanics are both reformulations of the mechanics of Newton that use different mathematical approaches, but yield exactly the same results (cf. Fließbach 2009). All three formulations can be shown to be logically equivalent.

Classical physics is generally thought to be deterministic. This means two things:

(1) Complete knowledge of the state of the system allows one to deduce all properties of the system and the results of any measurements taken.
(2) Complete knowledge of the state of a system in principle allows one to predict the state of the system at any future moment in time.

It is interesting to note that it follows that we can also predict all properties and results of measurements for the future.

Let us consider the non-relativistic case of two uncharged particles of masses m_1 and m_2 in free space.[5] Free space means that there are no external force fields. Or in layman's terms, there is nothing else to influence the movement of the particles. This example is known as the two-body problem. We assume full knowledge of the system. In this case this means, we know m_1 and m_2, the position of the particles as well as their relative velocities. From this knowledge, we can deduce all properties of the system such as momentum or energy. This corresponds to (1). The predictions of future states of the system, i.e. the future positions and velocities of the particles is also possible. In principle all we need for this are these two laws:

(1) The force acting on a particle is equal to its mass multiplied by its acceleration, i.e. rate of change of velocity.
(2) The law of gravity describing the magnitude and direction of the force acting on a particle in terms of the masses of the particle and the particle exerting the force as well as their relative position.

A solution of the two-body problem is presented in Fließbach (2009), and most other textbooks on classical mechanics. Strong doubts about the universal possibility of deterministic prediction in newtonian mechanics have been raised by Norton (2003). Norton suggests an example, known as Norton's dome, for which he claims the Newtonian equations have more than one solution. According to which of these solutions the system evolves would thus not be predictable. We will not discuss Norton's dome in any more detail here.

In classical statistical physics the system is not fully described, i.e. not all information about the state of the system, in statistical physics referred to as microstate, is given. Instead, the different possible microstates are assigned statistical weight. Such a description is referred to as a macrostate. The system is thought to be in a certain microstate; however, information about the precise state of a many particle system is impossible to attain and

---

5 Non-relativistic means we will not consider relativistic effects described by the theories of special and general relativity. This is a newtonian example, meaning it can be completely described by the mechanics of Isaac Newton.

usually irrelevant (cf. Fließbach 2010). The statistical weight assigned to the microstates is thus an epistemic probability. It is epistemic since it is not thought to be due to some uncertainty in the world, but only uncertainty of our knowledge of the world.

Lack of complete knowledge of the original state may have dramatic consequences in so-called chaotic systems (cf. Schurz 2006). Chaotic systems are systems for which a small difference in the original state may lead to very different later states. In a non-chaotic system a small difference in the original state will only lead to a very small difference in the consecutive states. This means that getting the original state slightly wrong will not mean the predictions will be completely off. In a chaotic system this doesn't apply. This can prove particularly problematic when the difference between two states is below what is detectable by measurement. If these two states will lead to completely different end states, accurate prediction is impossible. An example of a chaotic system is the double pendulum. Note, however, that the unpredictability of chaotic systems is still due to lack of knowledge about the world. In chaotic systems, this lack of knowledge simply has more dramatic consequences.

We already mentioned that in classical physics, complete knowledge of the state of a system allows an accurate prediction of any measurements taken. But what is measurement in classical physics? As we will see, measurement is a bit of an issue in quantum physics, so it makes sense to discuss what measurement is in classical physics. During measurement, the system under investigation is linked with a measurement device. The system and the measurement device interact according to the laws of classical physics. Ideally the effect of the measurement device on the physical quantity which we wish to measure is minimal. The system, however, has to influence the measurement device in a directly observable way that is somehow related to the quantity which we wish to measure. In electronic measurement devices the effect is usually translated into an electric signal which is in turn translated into a number on a screen. The following example was chosen because the device used is a bit simpler than that.

When using a mercury thermometer to measure the temperature of water, the thermometer has to be in contact with the water. This allows for exchange of energy between the water and the mercury in the thermometer until the temperature is in equilibrium, i.e. the temperature of the water and the mercury are the same. Since there is usually a larger amount of water than mercury, and due to the high specific heat capacity of water[6], the mercury will have little effect on the temperature of the water. The temperature of the mercury on the other side will adapt to whatever temperature the water has if given enough time. The temperature of the mercury is of course not directly visible, just as the temperature of the water isn't. Luckily the volume of the mercury correlates with the temperature though, so that we can simply check how much the mercury has expanded or contracted.

7. Experimental Basis of Quantum Physics

This section will discuss so-called double slit experiments, which are well-suited for explaining the necessity to replace classical notions with those of quantum physics. In order to understand the double slit experiments, one first has to understand the phenomena of diffraction and superposition. After explaining these we will consider the double slit experiment with light, which can be explained within classical physics. But, afterwards we will consider double slit experiments with other particles, whose results classical physics is unable to explain.

The diffraction of waves regards all waves, including for example water waves. The idea is that a when a wave front encounters a narrow opening in a barrier, roughly the size of the wavelength, it will move through the opening and behind the barrier it will spread out from the opening, but not just in the direction the wave originally moved in (Halliday et al. 2008, p. 963). Figure 1 shows an illustration of this.

Superposition is when the net effect of several effects occurring simultaneously is equal to the sum of both effects. One example of this is the superposition of waves which states that the displacement of two waves at any point in space will be added. This implies that two equal wave peaks will combine to a new peak, double the size of the original peaks.

---

6   The specific heat capacity is the amount of energy needed to raise the temperature of 1 kg of a substance by $1\,°C$

Figure 1: Diffraction of wave at a single slit. Image used with the permission of Philip Graemer.

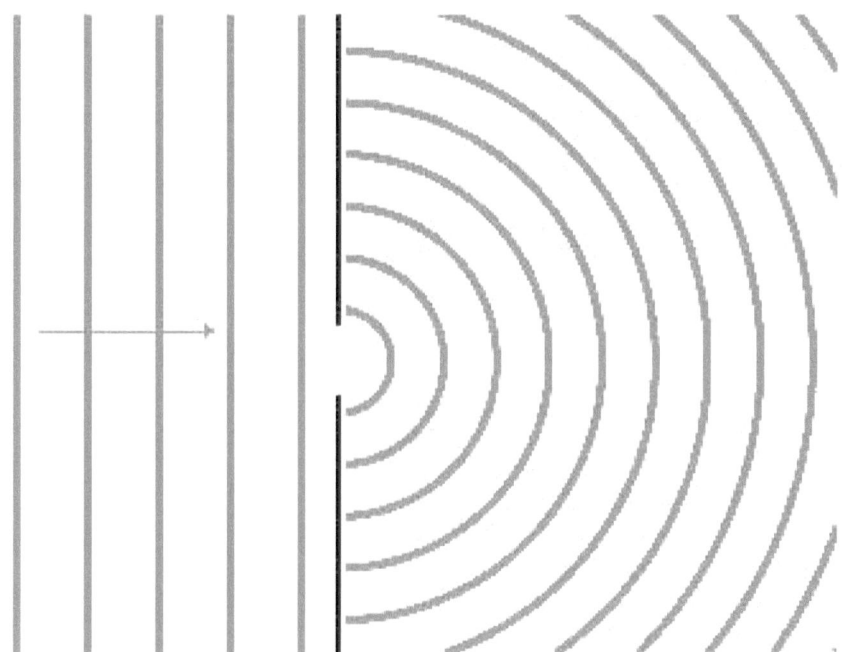

A wave peak and a wave trough on the other side will eradicate each other. These effects, called constructive and destructive interference, respectively are illustrated in figure 2.

The original double-slit experiment was used by Young (1802) to demonstrate that light is a wave. Two slits in a barrier are used to create two diffracted waves. They interfere with each other yielding an alternation of constructive and destructive interference between the waves from the two slits on a screen placed behind the barrier. This pattern has become known as an interference pattern. See figure 3 for an illustration of the experiment. The wave properties of light can be explained by classical electrodynamics. (cf. Jackson 2006, Fließbach 2008a)

Much more puzzling for classical physics are the following two versions of the double-slit experiment. The first one, done by Davisson and Germer in 1927, simply used a beam of electrons instead of light and managed to demonstrate that this will also yield an interference pattern.

In the second version, a single particle is send through the double-slit at a time. This is repeated several hundred or thousand times until one starts to see a pattern. One might expect that every particle will pass through one of the slits and be detected behind either one. The screen should thus detect particles as if half of them go through one slit and half through the other. This is, however, not the case. Instead, we still find the same interference pattern. This has been demonstrated for a number of particles, including photons[7], electrons, protons and neutrons (cf. Halliday et al. 2008, pp. 1068–1069).

Figure 2: Illustration of the principle of superposition showing constructive interference (top) and destructive interference (bottom). Image used with the permission of Philip Graemer.

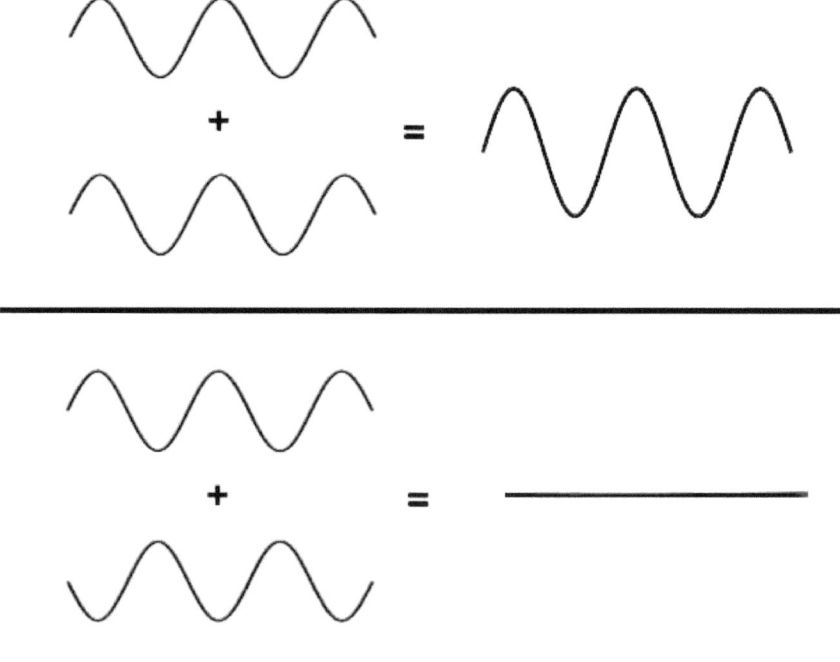

---

[7] Light particles.

Even more astonishing is the fact that the interference pattern vanishes if we detect through which slit the particle went. In this case we really do find the result we'd expect if half of the particles go through one slit, and half through the other (cf. Fließbach 2008b, p. 10–12). In the next chapter we will see how quantum physics explains these phenomena.

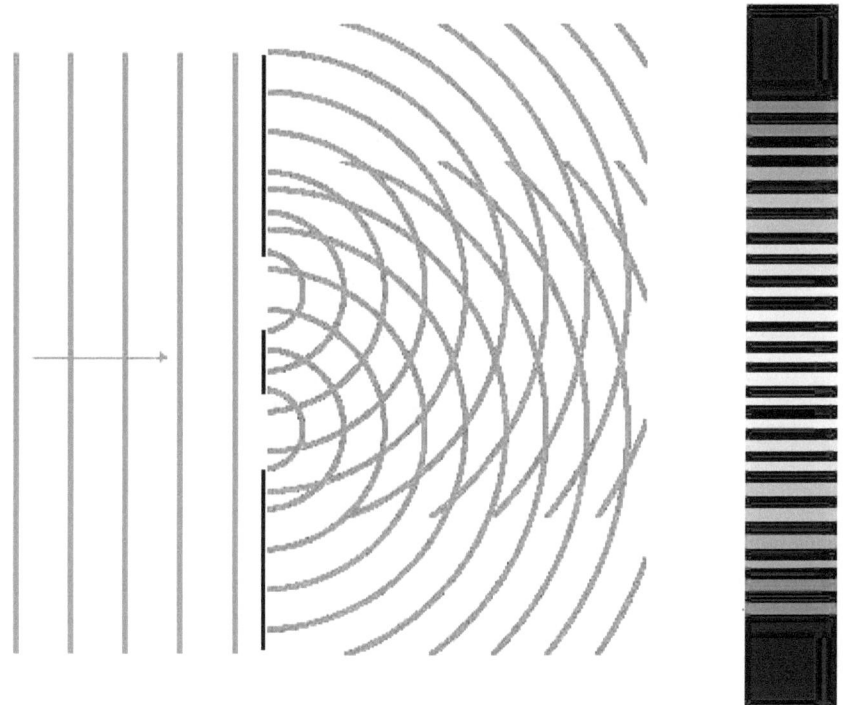

Figure 3: Scattering of a wave at a double slit showing the resulting interference pattern. Image used with the permission of Philip Graemer.

8. The Quantum Explanation

We will now discuss the basic explanation for these empirical results offered by quantum mechanics. There are of course different interpretations of quantum mechanics, some of which we will discuss later. However, this section will introduce the earliest attempts of an interpretation, on which the other attempts are based. First we will discuss the so called wave-particle duality and afterwards the Schrödinger equation.

8.1 Wave-Particle Duality

The 1927 experimental detection of interference in electrons goes back to an idea of de Broglie, expressed in 1924. Classical electrodynamics had treated light as an electromagnetic wave. The early 20th century, however, brought the discovery of the photoelectric effect and the Compton effect, that demonstrated that light had properties associated with particles such as momentum[8] (Fließbach 2008b, p. 5–9). De Broglie then suggested that if light is a particle, but also has wave characteristics, the same might be the case for other particles (Halliday et al. 2008, p. 1068). As we have seen, experimental results showed his intuition to be correct. Particles have a de Broglie wavelength λ, which is inversely proportional to its momentum (ibid.).[9]

This led to the notion of wave-particle duality. Originally being a wave and a particle were seen as contradictory. The former leads one to expect an interference pattern when performing a double slit experiment, the latter does not. According to the wave-particle duality, a particle moving through a double-slit will do so as a wave; however, when it is detected at the screen, this will be done as a particle.

We find this description a bit irritating, the problem being the characterisation of a physical entity as a 'wave'. We should rather consider the strict mathematical meaning of the term wave. A wave is nothing but a mathematical function that constitutes a solution to the 'wave equation'. Instead of asking ourselves in what way an electron is a wave, we should rather ask ourselves what physical attribute of an electron is described by a wave function. This is what we will do in the next chapter.

---

8 Momentum is mass times velocity.
9 $\lambda = h/p$, where h is the Plank Constant and p the momentum.

8.2 Schrödinger Equation and Born Rule

The Schrödinger equation

$i\hbar \frac{\partial \psi(q,t)}{\partial t} = H(q,p,t)\psi(q,t)$

is the basis of quantum physics, directly inferred from empirical evidence. We will not explain it in detail, but only give a rough idea of its meaning. The Hamilton operator $H(q, p, t)$, describes the system and $\psi(q, t)$ is the wave function. The wave function describes the state of the system and the Schrödinger equation how the state changes in time. Just like in classical physics the change of the system is deterministic and continuous, i.e. not abrupt. The state can interfere with itself, thus explaining the emergence of an interference pattern in the double slit experiment, even when a single particle is used.

But can it be explained that a particle is always detected at one location on the screen? It is, after all, only after several repetitions and addition of the individual detections that we start to see an interference pattern. From this we can deduce that it is the probability of detecting a particle at a certain place that is related to the interference pattern and thus the wave function. The exact relationship is described by the so-called Born rule.

While undergoing measurement, it seems that the state is changed from a superposition to one where the particle is localized. Unlike the evolution of a system as described by the Schrödinger equation, this evolution of a system undergoing measurement is discontinuous and indeterministic (Spekkens 2011). We thus have no way of predicting with certainty where a particle will be detected. We can only make probabilistic predictions. Unlike in classical statistical physics, this uncertainty is not the result of lack of information about the state of the system. Even when we have full knowledge of the state, we cannot predict the outcome. Some theories suggest that there exist hidden variables which would allow us to predict the out come. However, even these theories don't assume that we can gain knowledge of the hidden variables without measuring first. A detailed discussion of hidden variable theories will follow in the section on no-go theorems.

In section 6 we discussed measurement in classical physics. It seems that in quantum physics measurement plays a different role, since it influences the state of the system in a way not described by the Schrödinger equation.

The role of measurement in quantum physics has been the cause of great controversy. Penrose (2004) suggested that there is an unresolved empirical issue at hand, while Englert (2013) argues that the so-called measurement problem is a pseudo-problem and that no 'foundational matters are waiting to be settled' (Englert 2013, p. 1).

8.3 Observables

Physical quantities are properties of a system that can be quantified by measurement. In quantum physics, these are represented by *hermitian operators* in the so called *Hilbert Space*. Depending on the author, the physical quantities or the mathematical operators representing them are referred to as *observables* (Ismael 2009). The *eigenvalues* of the Operator correspond to the value of the physical quantity. The Born Rule, mentioned above, allows for statistical, but not deterministic, prediction of the outcome of measurement of a certain observable; while the wave function is *complex*, the values of observables are *real* numbers.

8.4 Einstein-Podolski-Rosen Paradox and Quantum Entanglement

In 1935 Einstein, Podolski and Rosen challenged the orthodox interpretation of quantum mechanics with what has become known as the Einstein-Podolski-Rosen Paradox (Einstein et al. 1935). In their minds quantum mechanics as understood at the time would require action at a distance, i.e. that an event at some location has an instant physical effect on events elsewhere. Special relativity, however, seemed to suggest that the effect of any event can only spread with the speed of light, but never instantaneously. The effect they described has become known as 'quantum entanglement' and has been empirically demonstrated. However, the interpretation offered by Einstein et al. is largely deemed impossible as we shall see in the next chapter. We will now introduce the concept of quantum entanglement and discuss its relevance.

Two interacting systems may be described by a single wave function, i.e. they are in fact described as a single system. The state of this system may be a superposition of two states. Both of these states are a combination of states of the subsystems. If both subsystems can either be in the state a or b the allowed combinations are (a, a), (a, b), (b, a) and (b, b). Let us now consider the case where the overall system is in a superposition of (a, a) and (b, b). Einstein et al. now considered that after a time of interaction the two subsystems are physically separated and one of them is measured. Since the overall system is in a superposition of (a, a) and (b, b), and during measurement the system will be reduced to one of the states that make up the superpositioned state we know that both subsystems will have to be in the same state. Measurement of one subsystem thus instantaneously influences the other subsystem to be in the state measured.[10]

This seems to contradict the notion of special relativity, that actions can only spread with the speed of light.[11] So Einstein et al. instead suggested that there are hidden variables which determine which state each subsystem is in before they are separated. They promoted what is described as a local hidden variable theory. The theory is local, because there is no action at a distance. It is a hidden variable theory, because the apparent indeterminism is believed to be the result of hidden variables, and not some objective chance in the world. John Stuart Bell is widely regarded as having demonstrated that such a local hidden variable is impossible. Bell's theorem and another so called no-go theorem will be the subject of the next section.

9. No-go Theorems

No-go theorems are theorems that forbid certain kinds of quantum physical theories. By making use of theoretical considerations, they demonstrate that some aspects a physical theory might have, would lead to predictions which contradict empirical results. We thus know that all theories that contain these aspects must be false. There are two important no-go theorems. The first one is Bell's theorem which states the impossibility of a local hidden variable theory which is consistent with the empirical facts (Bell 1964). The second theorem is the so-called Kochen-Specker theorem

---

10 Note that whether and along which lines one divides a system into subsystems is arbitrary. In the case of the Einstein-Podolski-Rosen Paradox the subsystems are chosen so that they are physically far apart in order to demonstrate action at a distance.
11 Since no information can be transmitted faster than light using entanglement the contradiction is only apparent.

which further limits the possibility of a hidden variable theory (Kochen and Specker 1967). We will discuss both these theorems without going in any detail about their proofs, which we deem too complex for this context.

9.1 Bell's Theorem

Bell (1964) builds on the Einstein-Podolski-Rosen paradox. He considers the case of particles with entangled 'spin'. Instead of performing the same measurement on both particles, he suggested measuring the spin in different directions and examine the correlations between the results. He was able to demonstrate, that any local hidden variable theory predicts that these correlations adhere to the so-called Bell inequality.

It has, however, been demonstrated empirically that in reality Bell's inequality is violated (Freedman et al. 1972). Instead, the results comply with the predictions made by orthodox quantum mechanics. It follows that any local hidden variable theory is empirically inadequate, as it cannot account for all facts which its alternative can account for. Since orthodox quantum mechanics also rejects locality, as demonstrated in the Einstein-Podolski-Rosen Paradox, this raises strong doubts, not just about a local hidden variable theory, but about locality in general.

9.2 Kochen-Specker Theorem

While Bell's theorem implies that a hidden variable theory must be non-local, the Kochen-Specker Theorem further limits the kinds of hidden variable theories compatible with quantum mechanics. In our discussion of the Kochen-Specker theorem we will largely follow Held (2013). At this point we will need to give a more precise statement of what we mean by a hidden variable theory. We understand a hidden variable theory to be one where value definiteness holds true. Value definiteness, as defined by Held (2013) is the following:

> '(VD) All observables defined for a QM system have definite values at all times.'
> (Held 2013, sec. 1)

This implies that, unlike presented in section 8, measurement doesn't change the values being measured, but it rather reveals values which exist previous to and independent of measurement. Measurement would thus have the same role it has in classical physics, which we discussed in section 6. Two further principles a quantum theory may or may not satisfy are the following:

If A, B and C are compatible observables, then:

(SR) If $C = A + B$, then $v(C) = v(A) + v(B)$
(PR) If $C = A * B$, then $v(C) = v(A) \cdot v(B)$

These are called the sum rule and the product rule respectively. *v(X)* refers to the value of the observable *X*. The sum rule thus states that the value of the sum of *A* and *B* is equal to the sum of the value of *A* and the value of *B*. The product rule states the equivalent for multiplication. Both of these seem to be plausible demands. They follow from a broader principle called the functional composition principle (FUNC), which states that the value of any function of an observable is equal to the value of the same function of the same observables, or in more formal terms:

(FUNC) 'Let *A* be a self-adjoint operator associated with observable A, let $f: \mathbb{R} \to \mathbb{R}$ be an arbitrary function, such that $f(A)$ is another self-adjoint operator, and let $|\psi\rangle$ be an arbitrary state; then $f(A)$ is associated uniquely with an observable $f(A)$ such that: $v(f(A))|\psi\rangle = f(v(A))|\psi\rangle$' (Held 2013, sec. 4)

The ψ indicates that the system is in an arbitrary state ψ. The Kochen-Specker theorem, however, states that compliance of (VD), as well as (PR) and (SR) contradicts orthodox quantum mechanics (cf. Kochen and Specker 1967 or Held 2013).

(KS) If quantum mechanics is true, the following two are contradictory:

(1) (VD) is true.
(2) (SR) and (PR) are true.

So for any hidden variable theory of quantum mechanics, (SR) or (PR) must be false. Of course this also implies that the general principle (FUNC) must be false. (FUNC) can, however, be deduced from a set of four principles. A hidden variable theory must thus reject at least one of these principles. The four principles are as follows:

(1) (STAT FUNC) 'Given $A, f, |\psi\rangle$ as defined in FUNC, then, for an arbitrary real number *b*:

$prob[v(f(A)) |\psi\rangle = b] = prob[f(v(A)) |\psi\rangle = b]$' (Held 2013)

(STAT FUNC) is a weaker, probabilistic version of (FUNC). A, f, ψ are the same as in (FUNC) and b is an arbitrary real number. It states that the probability of the value of $f(A)$ being b is the same as the probability of f of the value of A being b. (STAT FUNC) is implied by the formalism of quantum mechanics. This means that any quantum mechanical theory has to imply (STAT FUNC), unless it completely diverges from the original theory.

(2) (VD) Value Definiteness as defined above.

(3) (VR) Value Realism: 'If there is an operationally defined real number a, associated with a self-adjoint operator A and if, for a given state, the statistical algorithm of QM for A yields a real number β with $\beta = prob(v(A) = a)$, then there exists an observable A with value a.' (Held 2013, sec. 4)

Quantum mechanics allows us to calculate the probability of a value, given the state and the observable. Value Realism implies that if the probability calculated for a given operator and real value is real, then the operator represents an observable. Rejecting (VR) implies that construction of an operator from another operator which represents an observable doesn't automatically lead to another observable. An example would be the energy squared, which is constructed from the energy operator. Usually this would be an operator that represents an observable, same as energy on its own. If we reject value realism this isn't necessarily the case (Held 2013, sec. 5.2).

(4) (NC) Noncontextuality: 'If a QM system possesses a property (value of an observable), then it does so independently of any measurement context.' (Held 2013, sec. 4)

Measurement of a noncontextual property thus means determining a value which is independent of the fact that measurement is taking place and the way measurement is taking place. This is the case with classical measurement as described in section 6. Within a noncontextual theory, you can not make sense of a value of an observable without it being measured.

10. Interpretations

There is a variety of interpretations of quantum mechanics. They differ mainly in their position on the so-called measurement problem, but also in the theoretical entities they assume. The version of quantum physics we introduced, is close to the originally dominant interpretations. We will thus call it the 'orthodox interpretation'. The need for further interpretation derives from the obscurity when it comes to measurement and the sense of many that the nature of reality is not adequately described. One early interpretation was the so-called Copenhagen Interpretation, which is still popular today. The issue has, however, been the focus point of so much debate, that there are now actually several interpretations of this interpretation.

This section will give an overview of some of the many attempts to interpret quantum physics, but before we introduce these interpretations, we will discuss the famous Schrödinger's cat thought experiment, which originated much of the controversy regarding measurement. The different interpretations can best be explained in regard to their position on the thought experiment.

10.1 Schrödinger's cat

Schrödinger's cat is a thought experiment, suggested by Erwin Schrödinger in 1935 to illustrate absurd consequences of the orthodox interpretation of quantum physics (Schrödinger 1935). The experiment requires a 'quantum box', which shields the inside from interaction with the outside in order to create an isolated system. Inside the box is a radioactive source. At what point such a source emits radiation depends on quantum mechanical processes of chance. The source can thus be used as a generator of chance of sort. But as long as the system is not measured, i.e. the box is opened, the source has to be seen as being in a superposition state of having emitted radiation and not having emitted radiation. Schrödinger further imagines a Geiger counter to be inside the box, which detects whether the radiation was emitted or not. The system of radiation source and Geiger counter is thus in a superposition state of 'radiation emitted and radiation detected' and 'radiation not emitted and radiation not detected'. The detector is connected to an apparatus in such a way that it breaks a glass with poison if radiation is detected. Additionally, there is a cat inside the box which dies if the poison is released. The whole system is thus in a superposition of two states, in one of which the cat is dead and the other alive. We can never observe this state though, since once we open the box, the state will be reduced to one of the two options.

Is such a superposition of macroscopic objects, like cats, really possible? The thought experiment raises the question, at what point a superpositioned state is *measured* and thus reduced to the measured state, if this is what happens at all. One possible answers is that the detector is already carrying out a measurement. A very strange possible answer is that the cat is measuring, simply because it is a conscious being. Note that in principle, one could also replace the cat with a human being. If measurement and reduction is taking place within the box, this certainly demands an answer to the question, which precise process or interaction constitutes such a measurement?

10.2 Subjectivist Interpretations

Subjectivist interpretations view quantum physics as a tool to predict subjective probability of measurement. In the case of Schrödinger's cat, this tool can be used from the perspective of the Geiger counter, the Cat or the person who opens the box. The probabilities will be the same in all cases, so there will be no conflict of results. If the probability of radiation being emitted is 50%, then there will be a 50% chance for the detector to detect radiation, a 50% chance for the cat to inhale to poison and a 50% chance for the person who opens to box to find the cat dead. So when we consider a measurement to be taking place has no empirical relevance. Some versions of the Copenhagen Interpretation are subjectivist.

10.3 Objective-Collapse Interpretations

Objective-Collapse Interpretations hold that there is an objective state of the system, which collapses when measured. Collapse theories have to tell us at which point a superpositioned system, such as Schrödinger's cat, collapses. Penrose (2004) argues for an objective-collapse interpretation. He suggests an experiment, called FELIX, which arguably can not be accounted for by subjectivist interpretations. In it different results will be measured, depending on whether a mirror, which is part of the set up, can be in superposition or not. Penrose argues that whether a state collapses is related to the gravitational potential.

10.4 Modal Interpretations

Modal Interpretations, defended among others by Van Fraasen (1991), hold that the description of the state through the Schrödinger equation is complete. There is no state collapse or reduction, neither in the subjective, nor objective sense. Instead the system has a definite state at all times, which evolves deterministically in accordance with the Schrödinger equation, including all cases of interaction with particles, which are not part of the system. It is merely the values of the observables which are determined by chance. The possibilities of the values of the observables and their assigned probabilities are thus implied by the state, hence the name *modal* interpretation.

10.5 Many-Worlds Interpretation

The Many-Worlds Interpretation doesn't only offer a view on the evolution of states, but also interprets them in a certain way (cf. Vaidman 2014). According to this Interpretation, first suggested by Everett (1957), the universe contains many different worlds. Each time a quantum chance event takes place, new worlds are created, one for each possible outcome. Superpositioned states constitute certain interactions between the worlds. So with regard to Schrödinger's cat, there are two worlds, one in which the cat is alive and one in which it is dead. In the double slit experiment there is one world in which the particle goes through the left slit, and one in which it goes through the right slit. However, these two worlds are still linked in a way, which explains the possibility of superposition events.

10.6 Hidden-Variable Interpretations

A hidden-variable interpretation is any interpretation, which postulates unknown variables, which determine the apparently indeterministic processes in quantum physics. The hidden variables are not only unknown, but unknowable before measurement. Hidden-variable interpretations thus don't allow us to make additional predictions about measurement results. It is only after measurement we can decide which hidden state the system was previously in. The apparent randomness is thus thought to be due to epistemic chance as in classical statistical physics, and not objective chance. As opposed to classical statistical physics complete knowledge of the state is not only impossible for technical reasons but in principle.

As discussed in section 9, the possibility of hidden variable theories is limited by so-called no-go theorems. The local hidden variable theory imagined by Einstein et al. (1935) is forbidden by Bell's theorem. An interpretation which is not forbidden by any no-go theorem is the de Broglie-Bohm theory (cf. Goldstein 2013).

As in the orthodox interpretation, the de Broglie-Bohm theory includes a wave function governed by the Schrödinger equation. Unlike in the orthodox interpretation a full description of the system also includes an actual position of the particle. The movement of the particle is described by an additional equation, called the guiding equation, which depends on the wave function. One could say that the wave function guides the particle. In a double slit experiment with a single particle, the outcome is already determined by the original position of the particle. However, this position can only be known in hindsight, i.e. after the experiment has been performed.

10.7 Operational Approach

While all interpretations discussed so far included a description of a *state* of a *system*, the operational approach doesn't intend to describe such a state at all (cf. Spekkens 2011). Instead the aim is to devise rules which allow to predict the results of measurements based on the experimental set up. The experiment is divided up into the *preperation* and the *measurement*. The former describes the way the system is prepared and the second the way it is measured, i.e. the measurement set up. Every action on the system has to be described as part of the preparation or the measurement. The goal is now to simply find rules which allow (probabilistic) prediction of measurement results, based on knowledge of preparation and measurement set up.

Part III. Reflections on Quantum Physics

In this part, we will use some of the philosophical foundations laid in part 1 to reflect on quantum physics and its various interpretations. First, we will discuss various questions related to theory and observation in quantum physics. Then, we will ask whether quantum physics is acceptable, despite various 'odd' characteristics. Finally, we will discuss ontological questions related to quantum physics. We will suggest that some of these thoughts may be relevant to attempts to combine quantum physics with gravitational theory.

11. Theory and Observation in Quantum Physics

11.1 Observables and Observation

Their name may suggest that 'observables' are observational terms, but is this really the case? As we discussed in section 8.3, observables are measurable physical quantities. Does this qualify them as observations? The answer is clearly no, since measurement and observation are not the same. We shall discuss the difference, as well as the relationship between measurement and observation and thereby establish the status of observables as theoretical terms, which are, however, more directly linked to observations than other theoretical terms.

When a physical system undergoes measurement, it interacts with a measuring device, which changes in a way that is linked to the aspect of the system being measured. The change of the measuring device must be observable. We explicated this classical notion of measurement in section 6. It is clear that the aspect of the system being measured, which in quantum mechanics is the observable, is not itself content of our observation. Physicists will in general observe a graphic or numerical representation of the measurement results on a computer screen, but certainly not the observable directly.

Another way to demonstrate that measurement is not observation is through the criterion of ostensive learnability. The terms used to describe measurement results are not ostensively learnable. There is no ostensive learning experiment that has any chance of teaching someone the meaning of 'momentum' or what it means for an unobservable particle to have the momentum 1 Ns.[12]

We have, however, seen that measurement seems to play a slightly different role in quantum physics, a role which is still the subject of controversy. Two hypothetical differences we should consider are the following:

(a) Measurement has an effect on the physical system which is not in accordance with its normal evolution.
(b) The observables have no values independent of measurement.

The different interpretations we learned about in section 10 imply different views on whether these are true. (b) constitutes a rejection of value definiteness, which we will discuss again in section 12.2. For now, all that needs to concern us is whether (a) or (b) change anything about the status of observables as theoretical terms. That this is not the case doesn't require great explanation. The reasons given above for observables not being observational terms are independent of (a) and (b).

Even though observables are not observational terms, they are nonetheless linked more closely to observation than other theoretical constructs in quantum physics, such as the wave function. Observables can be measured and measurement can result in some observable change in the measurement apparatus, or a device linked to it such as a computer screen. So, the link from observables to observation is relatively brief. The wave function, on the other hand, can only be used to predict the probability of measuring certain values of observables. It is thus one step further away from observation.

11.2 Theory or Interpretation?

In section 10 we referred to the different versions of quantum physics as interpretations. But when do different interpretations of a theory stop being mere interpretations and become separate theories? The answer given by Van Fraasen (1991, pp. 241–244) is that two different theories make different empirical predictions, while two interpretations of the same theory agree on the empirical predictions, but disagree on matters which are beyond empirical phenomena. One might say that the interpretation describes a

---

12 Ns is Newton seconds, the unit of momentum.

structure of reality which is in accordance with the empirical predictions of the theory. We will distinguish this matter further in the next section. What is important is that the interpretation makes a stronger claim than the theory by itself.

Van Fraasen (1991, p. 243) further claims that 'attempts to interpret are very much like, if not the same thing as, attempts to introduce hidden variables'. The idea is that the additional content of an interpretation is beyond the empirical phenomena and thus a *hidden variable*. Hidden variable theories are, however, a much more specific kind of theory. As introduced in section 10.6, they attempt to explain the apparently indeterministic nature of quantum mechanics through the existence of definite hidden variables. Knowledge of these variables, although not necessarily believed to be accessible, would in principle allow for deterministic prediction. But not all interpretations assume determinism. The collapse theory and modal theory have a theoretical structure which goes beyond empirical phenomena, including for example the wave function. They are nonetheless indeterministic and don't assume value definiteness.

Penrose (2004, pp. 782–784), for example, is unsatisfied with mainstream views on quantum mechanics, because they tell us 'essentially nothing about an actual *quantum reality* of the world' (Penrose 2004, p. 782). But the description of reality offered by him includes an indeterministic evolution of the quantum state. This is in no way contradictory to his realism.

The distinction between theory and interpretation offered by Van Fraasen is certainly possible. It does, however, imply that two interpretations of the same theory may have a completely different and contradictory theoretical structure. It may well be argued that this should be sufficient to classify them as distinct theories. Van Fraasen himself seems to do this in 'The Scientific Image' (Van Fraasen 1980, pp. 10–11). In the end this boils down to pure semantics. For this reason, we don't pay particular attention to when we use the word 'theory' and when 'interpretation'. What is important is that according to the realist and literal reading of theories, two theories or interpretation are significantly distinct from one an other if they have contradictory theoretical structures. Whether one classifies them as opposing theories or merely interpretations of the same theory is a matter of choice.

There is, however, a major problem with classifying some of the views on quantum physics. They hold different views on the *measurement problem*.

This means that they disagree on something which is neither an empirical matter, nor a matter of *description of reality*. The disagreement is about whether the orthodox view of quantum physics is complete or not. Someone in favour of the collapse view may for example hold that the subjectivist view is incomplete, because there are empirically relevant cases for which the subjectivist can not account. The subjectivist would disagree on the existence of such cases. The point of the collapse theorist would not be that there is empirical evidence contrary to the subjectivist view, it is rather that there supposedly are cases for which it is nor clear which empirical predictions follow from the subjectivist interpretation.

11.3 Theory and Ontology

We believe that the dichotomy of *empirical theory* and *realist interpretation* requires further differentiation. As we have argued through out part 1, the theoretical structure of a theory doesn't have to be interpreted in the realist sense. But neither is it simply observation. So, there is room for interpretation which is not occupied with a description of the 'real' world. We suggest that, besides the empirical aspect of quantum theory, one should recognise an ontology as well as a theoretical structure.

In section 5 we introduced scientific realism as the notion that theoretical entities refer to real (mind-independent) objects. So, scientific realists believe that the theoretical structure represents some aspect of reality. This is the ontology. For a scientific realist, theoretical structure and ontology are thus closely linked. We think that a distinction is necessary anyway, for two reasons. The first is that not everyone accepts scientific realism. The second is that scientific realists who agree upon a theoretical structure, may still hold opposing views on what aspects of it are representations of reality and how they represent it.

The interpretations of quantum physics we discussed imply different theoretical structures. Most notably they differ on how the wave function evolves and how it is linked to observables. In the collapse interpretation the wave function collapses at some point, while in the modal interpretation it evolves continuously. During collapse the wave changes in an indeterministic way. The resulting state of the system yields the value of a measured observable. In the modal interpretation the wave function doesn't change in an indeterministic manner, but the value of an observable can not be deterministically predicted by the wave function.

Proponents of the many-worlds interpretation suggest that the superposed states correspond to different worlds and that all possible outcomes, even those not realized in our world, are real in some world. This should be seen as a question of ontology. It clearly surpasses a mere description of the evolution of the quantum state. It also includes a description of the reality which is believed to underlie the theoretical structure.

11.4 Structural Correspondences of Interpretations

It is not entirely clear if all the interpretations we discussed make the same empirical predictions under all possible circumstances. But they do agree on a wide range of empirical matters. Otherwise it would be a matter of a few simple experiments to falsify some of them.

Section 5.6.3 discussed intertheoretical correspondences. It seems plausible that the interpretations of quantum physics, which agree upon a wide range of empirical phenomena, should have some theoretical correspondences between them. It turns out that the wave function and the Schrödinger equation play a central role in all the interpretations we discussed. The interpretations vary in two aspects:

(a) Under which conditions, if any, the wave function evolves divergent of the Schrödinger equation.
(b) How the wave function is related to observables.

Disregarding these differences, the wave function is still an essential aspect of all these interpretations. It is fair to say that this intertheoretical correspondence explains the shared empirical success. Phenomena such as the interference pattern in the double slit experiment are typical wave phenomena. Therefore, it comes as little surprise that a wave function plays a key role in many if not all theories that are able to explain these phenomena.

12. Is Quantum Physics Acceptable?

Quantum physics is able to explain a wide range of empirical phenomena. However, it brings with it some strange consequences such as indeterminism and arguably also the rejection of value definiteness. In addition, some commentators on the measurement problem suggest that quantum physics is incomplete. This raises the question whether quantum physics is an acceptable theory. We will discuss the three mentioned issues in turn.

12.1 Indeterminism

Newtonian physics was arguably one of the greatest advances in the physical sciences. No previous theory was as all-embracing as the physics of Newton. The movement of all objects could be understood as compliant with a very limited and fairly simple set of deterministic laws. Newton impacted the understanding of science itself to such an extent that science arguably became synonymous with explanation through deterministic laws. While this effect of Newtonian theory on our thinking about science is understandable, it is only partially justified. Explanation and prediction of individual phenomena through more abstract laws is undoubtedly a key aspect of scientific methodology, but it is by no means necessary that these laws are deterministic. Quantum physics is an example for this.

Without a certain uniformity of the world, science is impossible, since we would be unable to derive empirically adequate, abstract laws. But a *deterministic uniformity*, in other words a uniformity, not just of probabilities, but of actual phenomena, is by no means necessary. Probabilistic laws are not as rich in content as deterministic laws, since they don't contain certain, positive information about what will happen under given circumstances. But they have content nonetheless and thus do constitute substantial insight.

That there are uniformities in the world at all is not given a priori, since it is not logically necessary. So we should count ourselves lucky that science is as successful as it is. If we now also demand that observable phenomena be explainable by deterministic laws, that is just asking for way more than we can expect. It would no doubt be convenient if this was possible. But something being convenient doesn't necessarily make it true. Newtonian physics gave us too high expectations of what science can achieve. Now we have to lower them.

This doesn't mean that we shouldn't look for deterministic explanations. They are after all richer in content and finding them would thus constitute an advancement. We just shouldn't be surprised if we don't find deterministic explanations. There is no necessity for them to exist. It is likely that for quantum physics the search is futile. Even the so-called hidden variable theories don't assume that we can ever make deterministic predictions about quantum phenomena. They only assume that the underlying theoretical structure or reality is deterministic.

12.2 Value Definiteness

We have seen in section 9 that there are strong problems when trying to construct a version of quantum physics in which value definiteness holds true. There are some hidden-variable theories like the De Broglie-Bohm theory, which are not contradictory to the Kochen-Specker theorem. But nonetheless, we have to consider the possibility of abandoning value definiteness. In this section, we will try to get a clearer picture of what this implies and whether a version of quantum physics where it doesn't hold true could be acceptable.

An account of a hypothetical measurement which would force us to either reject value definiteness or basic principles of logic is given by Van Fraasen (1991, p. 107–108). Imagine there are three observables, which we can measure in principle. Let us call them A, B and C. In quantum physics it is common that certain observables can't be measured simultaneously. Let us assume that in this case we can measure any two of the three observables at once, but never all three at once. For all three observables we can expect any value from 0 to 1. We repeat the experiment several times for each pair of observables. Consider now we find that for the pair A and B we always find that in half of the cases the measured value of A is 0 and that of B is 1. In the other half of the cases it is the other way around, i.e. A has the value 1 and B has the value 0. We can now make clear probabilistic predictions about the results. Now let us say that for the other two pairs we find exactly the same results. We can derive the following two statements from our findings:

(a) 'No two of A, B, C have the same value.'
(b) 'Each of A, B, C has value 0 or 1.' (Van Fraasen 1991, p. 107)

Clearly these two statements are contradictory. If A, B and C all have either value 0 or 1, then it is necessary that at least two of them share the same value. So how do we deal with such a contradiction?

The most obvious response would be to say that such a scenario, since it implies a logical contradiction, cannot exist. Since this particular scenario was made up by Van Fraasen and doesn't constitute a real experiment, there is no problem. There are, however, similar, real cases discussed in the literature.

It also seems counter-intuitive to think that a series of measurement results can ever constitute a logical contradiction. What logical principle could ever forbid certain measurement results in a series of experiments? Certainly this is not possible. It is only our interpretation of the results that can be illogical. So, instead of rejecting classical logic and trying to find non-classical logics to replace it (as some authors do), we should re-examine the hidden assumptions that led us to postulate (a) and (b), which, without doubt, really are contradictory.

Van Fraasen identifies two hidden assumptions that leads us to (a) and (b). The first is value definiteness, i.e. that A, B and C have values independent of measurement. The second is that measurement faithfully reveals those values. Let us have a brief look at the latter first. If A, B and C have values independent of measurement, but these are not revealed by measurement, then it can't really be considered measurement of A, B and C. What ever else this process might reveal – it is not a measurement of A, B and C. We might instead call the quantities we measure A', B' and C'. But now we have just moved the problem to these new observables. The statements (a) and (b) now hold true for them. So rejecting the second principle without also rejecting value definiteness offers no real solution.

We thus have to choose between basic principles of classical logic and value definiteness. Van Fraasen chooses logic. But is a rejection of value definiteness acceptable? In the case of the double slit experiment, discussed in section 7, this implies that if we don't measure through which slit the particle goes it doesn't have any location at all. But why should it always have a location? This may be contrary to the thinking about location in classical physics, which is also a lot closer to the way we think about it in our every day life. But it is by no means contrary to scientific reasoning. As part of a scientific theory we may well postulate quantities which only have values when measured. This still allows for (probabilistic) prediction of the outcomes of measurements. In section 13.1 we will argue that it is also compatible with the notion of realism.

12.3 Measurement Problem

While we should have no problem with indeterminism or the rejection of value definiteness, the question remains whether quantum physics is complete. So, the question is not if quantum physics has some consequences which might be considered undesirable, such as indeterminism or the rejection of value definiteness. Instead we have to ask whether quantum theory can stand by itself as a theory with wide ranging empirical consequences.

What is perceived as a problem by some, is the lack of a clear description of the process of measurement in quantum physics. The two main issues appear to be:

(a) Does *measurement* influence the quantum mechanical state in a way that is not described by the Schrödinger Equation?
(b) Under what conditions does such a *measurement* occur?

Penrose (2004, pp. 856–865) suggests the so-called FELIX experiment to test his hypothesis on how gravity influences a collapse of the wave function. To our knowledge, such an experiment has not been conducted to this day, due to difficulties in its realisation. The basic idea is 'to construct a 'Schrödinger's cat' that consists of a tiny mirror M, placed in a quantum superposition of two slightly different locations' (Penrose 2004, p. 856). The experimental set up is supposed to ensure that a superposed mirror and a *collapsed* mirror will yield different measurements of light reflected from the mirror. Penrose gives a more detailed description of the experiment and suggests that the outcome depends on the gravitational potential of the mirror. Like in the Schrödinger's cat thought experiment, a macroscopic superposition is induced. The difference is that in this case we should be able to experimentally verify whether the object was in superposition or not and thus at what point a 'collapse' of the wave function takes place. This contradicts the notion held by proponents of the subjectivist interpretation (cf. section 10.2) that it does not matter at what point 'measurement' takes place.

Others, like Englert (2013), argue that there aren't really any problems with standard quantum theory. No fundamental issues need to be solved. We are unable to resolve this matter here and now, but it is fair to say that even if their is no real problem, at least there are some issues that need to be made clearer in order for everyone to accept it as a pseudo-problem.

13. Ontology of Quantum Physics

In this section, we will discuss the ontological status of quantum theory. We will first discuss what implications the rejection of value definiteness would have for a realist or literal view on quantum physics. Afterwards, we will pay attention to the different interpretations introduced in section 10. We will be asking how compatible they are with the different ontological viewpoints, including realism, instrumentalism and the position advanced by us in section 5.7. According to our view scientific theories shouldn't be considered a descriptions of reality, but should nonetheless be taken literally. In the final subsection we will argue that such literalness is potentially important for resolving unsolved issues in quantum physics, such as the unification of quantum physics and gravity.

13.1 Value Definiteness and Ontology

Does the rejection of value definiteness imply a rejection of realism? Arguably observables can not reference real properties of real objects if they don't have values independent of measurement. There are still the possibilities that either the real objects only have the properties during measurement, or that the objects themselves only exist during measurement. Both views are not necessarily contradictory to realism. This would, however, be a very limited form of realism, which doesn't correspond to what most realists have in mind. After all, a realist wants to discover an independent reality, and not one that is created through measurement.

A more fruitful way to save realism is by saying that not all theoretical terms and entities need to have a real reference. Just because certain theoretical entities, such as observables don't have real reference, doesn't mean that others can't have them either. We think the most obvious candidate for real reference is the wave state. A quantum mechanical system is always in a definite state, even if this state doesn't include a definite location or momentum. So, one may well take this theoretical state to correspond to an actual state of reality. Realists have to accept that quantum reality looks very different to the reality pictured by classical physics, and that terms like location and momentum are not a part of it or at least play a very different role. Accepting this should be easy for constructive realists, who already limit their realism to certain structural correspondences between theory and reality.

We have argued in section 5.6.5 that realism should be rejected because it adds no empirical content. Theories should nonetheless be taken literally in order to allow for further research, based on such a literal reading. For the same reason realists should take the wave state as corresponding to reality, we think the wave state is what should be taken literally. The difference between our view and realism is, as always, that we don't take this state to reference reality. The literal interpretation of the wave state serves as a basis for further research. The matter of open issues in physics will be discussed further in subsection 13.3.

As we have seen, value definiteness is not necessary for a constructive realist reading of quantum theory. It seems that resistance to the rejection of value definiteness might be motivated by a naive understanding of realism. According to naive realism, reality is exactly as we experience it (cf. section 5.2.1). The values of observables measured by us, one might follow, thus also have to be part of reality.

It is, however, questionable whether value definiteness actually follows even from a naive form of realism. As we have argued in section 11.1, statements about the values of observables are not actually statements about observation. They are in fact theoretical statements, even though they are more directly linked to observation than the wave state. So, observables are not observations. It follows that, while naive realism might be a motivation for defending value definiteness, it is not a justification of it. Independent values are not required by naive realism, since these values are not observations.

Even if naive realism required a commitment to value definiteness from us, the problem quantum mechanics poses for value definiteness remains. If some forms of realism really require us to hang onto value definiteness, maybe it is time to reject such forms of realism. Besides various anti-realist views, more sophisticated forms of realism, such as constructive realism pose an alternative. Modern science, and quantum mechanics in particular, demand an ontology which allows for a more complex reality. This is particularly so since observations are often, especially in quantum mechanics, very indirect. If we want any part of quantum mechanics to be real, then reality certainly needs to have a more complex relationship to observation than a simple 'reality is just as we perceive it'.

13.2 Interpretations and Ontology

This section will deal with the different interpretations of quantum mechanics and their compatibility with the various ontological view points. Our ontological views may guide which interpretations we deem acceptable. If they are incompatible we may reject the interpretation of quantum mechanics a priori. We will, however, argue that most of the interpretations are largely ontologically neutral.

13.2.1 *Subjectivist Interpretations*

The subjectivist interpretations can be seen as examples of logical positivism or instrumentalism. At what point *measurement* takes place is deemed irrelevant on the ground that it changes nothing about the probabilistic prediction. The subjectivist interpretation thus serves fine from an instrumentalist point of view. A literal or realist reading on the other hand seems at first sight impossible, since it doesn't offer a single state, which can be said to be the state of the system. Instead there are several states, depending on the point we consider *measurement* to be taking place.

We believe that it should, however, be possible to construct a single state from the many states in the subjectivist interpretation. One way to do this would be to form an ensemble of all the individual states. While none of the individual states can be said to be more significant than the others, we could read this ensemble literally or even consider it to correspond to reality in some way. The ensemble would include all the individual states and thus all relevant information. Such an ensemble is not included in the classical subjectivist interpretation, but this goes to show that the subjectivist interpretation is not in principle contradictory to a literal or realist reading. We believe that the real conflict between the subjectivist interpretation and the objective-collapse interpretation concerns their different takings on the measurement problem, not ontology. This doesn't mean that they don't imply very different descriptions of reality. But the subjectivist interpretation is not inherently linked to a rejection of realism. We will argue in the next section that such an ontological neutrality holds true for the objective collapse interpretation, too.

13.2.2 Objective-Collapse Interpretations

The objective-collapse interpretations are often regarded as realist interpretations. In fact, one of the main motivations for proposing an objective-collapse interpretation cited by Penrose (2004, p. 782 ff.) is that he is unsatisfied with the lack of a clear description of reality of other interpretations. But just because the objective-collapse interpretation implies a fairly clear description of a state, doesn't mean it presupposes realism. We can accept its theoretical structure and empirical consequences without believing that this theoretical structure references an independent reality. It is thus also ontologically neutral. We can take the theoretical structure literally, without implying realism or we can accept it as a useful instrument for empirical prediction. This doesn't even need to imply a literal reading. While realism does require a proper description of reality, instrumentalism doesn't necessarily require a lack of such a description. It can always treat this description as a framework for prediction, without associating any larger meaning with it. A literal reading goes slightly further in the direction of realism, because it doesn't treat two different theoretical structures with identical empirical content as equal.

An instrumentalist acceptance of the objective-collapse interpretation thus accepts it as equally valid to all other interpretations which are empirically equivalent. It may be favoured or disfavoured for pragmatic reasons such as its comprehensibleness and difficulty of calculation in its mathematical framework. A literal acceptance also implies a commitment to the objective-collapse interpretation as a basis for further research.

A realist or literal reading of the objective-collapse interpretation implies realness or literalness of the wave function, since the wave function is seen as *objectively* collapsing. We have endorsed the view that the wave state should be taken literally in section 13.1 and thus see this as a positive aspect of the objective-collapse interpretation. Whether one should favour an objective-collapse interpretation over other interpretations which allow to take the wave state literally does, however, ultimately depend on what one makes of its taking on the measurement problem. Penrose (2004, p. 816 ff.)

offers a very specific view on the conditions of collapse and empirical ways to test it. Acceptance of his interpretation can thus not solely be based on ontological considerations.

13.2.3 Modal Interpretations

Modal interpretations allow for a realist or literal reading of the wave function. Unlike with objective-collapse interpretations, the wave function is not subject to collapse, but behaves completely in accordance with the Schrödinger equation. So, there are important theoretical differences between modal and objective-collapse interpretations. Observables don't correspond deterministically to certain wave states and can thus be said to be completely distinct theoretical entities. The possibilities of ontological interpretation are, however, essentially the same. The wave function can be regarded as corresponding with reality; it can be taken literally without implying any connection with a mind-independent reality, or the whole theoretical structure can be seen as a mere instrument for prediction. Van Fraasen (1991, p. 273 ff.) is one proponent of a modal interpretation. He is also a defender of a literal, but not realist reading of scientific theories (cf. Van Fraasen 1980). A realist or instrumentalist reading of the modal interpretation is, however, also possible.

13.2.4 Many-Worlds Interpretation

In section 11.3 we differentiated between theory and ontology. This distinction is particularly important with regard to the many-worlds interpretation. The many-worlds interpretation, besides having a particular theoretical structure, postulates that certain elements of that structure constitute different worlds. This is the ontology. For the theoretical structure, one might in principle say the same, we have said for the objective-collapse and modal interpretations, i.e. that it can be read in a realist, literal or instrumentalist manner. The idea that there really are different worlds is, however, clearly a realist notion. It may in principle be possible to have a realist reading of the same theoretical structure, which doesn't imply such worlds. But the postulate of many-worlds is closely associated with this interpretation, which explains its name.

Besides our general criticism of scientific realism expressed in section 5.6.5, we think that the ontology of the many-worlds interpretation is particularly speculative. Constructive realists limit themselves to postulating certain structural correspondences between theories and reality. But the many-worlds interpretation goes so far as to postulate the existence of multiple worlds. This seems to be a quite extensive ontology with no, or at least very limited, empirical consequences. It is highly doubtful that proponents will ever be able to present sufficient justification for it.

13.2.5 Hidden-Variable Interpretations

The de Broglie-Bohm theory is praised by Penrose as being 'much more down to Earth' (Penrose 2004, p. 789) than other interpretations. What he means is that it is much closer to the kind of clear description of reality, which he aspires to. He perceives a weakness in the existence of '*two levels of reality*' (ibid.), one being the particle with an actual position, the other being the guiding wave. There is a parallel to the modal interpretation, which treats the wave function and the observable as two separate entities. Penrose deems such a splitting of reality unnecessary. As with the other interpretations a literal or instrumentalist reading is, however, also possible.

The defining characteristic of all hidden-variable interpretations is that all quantities are considered to have definite values at all times and that evolution of the state and measurement outcome is guided by deterministic laws. As we have argued in section 12.1, these aspects are, however, not necessary for scientific realism or science in general. Determinism and value definiteness can thus not be used to justify hidden-variable theories over other interpretations of quantum physics.

13.2.6 Operational Approach

For all interpretations so far, we have argued that they are to a large extent ontologically neutral. There is, however, a decisive difference when it comes to operational approaches. The idea of an operational approach is that it doesn't yield a description of a state of a system. Instead, it merely offers a way to calculate measurement outcome based on the experimental preparation. This is a very instrumentalist approach, that doesn't allow for a realist or literal reading. If there is no state, but only measurement outcomes, then there is nothing which can be said to correspond with reality or be taken literally. While as we have argued, a description of a theoretical state doesn't necessarily require a realist or literal reading of that state, the absence of a state does imply instrumentalism. Even a realist may adopt an operational approach to calculate measurement outcomes. But he can never be satisfied by an operational approach alone, because it says nothing about the structure of reality. In the same way there is nothing to be taken literally, on which further research could be based. Operational approaches, one might argue, correspond to crude collection of empirical facts and have nothing to do with the abstract theorizing so crucial for the advancement of the sciences.

13.2.7 Interpretations and Ontological Parsimony

We have argued that most interpretations are compatible with all ontological positions. It follows that we can not dismiss these interpretations based on an incompatibility with whatever ontological position we deem to be right. Ontology may however still have something to say about choice of interpretation if we aspire ontological parsimony. The problem with ontological parsimony is that it is often difficulty to decide which theory is more parsimonious. It is easy to reject theories with entirely superfluous elements in favour of a reduced version of the theory. But this is a very specific case and it is not applicable if there are two theoretical structures or ontologies which are simply different, without one being a simpler version of the other. So, ontological parsimony can be of very limited use for choosing an interpretation of quantum physics. There are however two arguments which might be applied:

(1) The many-worlds interpretations may be rejected on the basis that many worlds are more extensive than one world.
(2) Penrose (Ibid.) argued that hidden-variable interpretations are more extensive than the collapse interpretation because they contain two states; the state of the particle and the guiding wave. This argument may also be applied to the modal interpretation because observables are not deterministically linked to a wave state and thus, one might argue, constitute separate entities.

13.3 Literalness and the Challenge of Quantum Gravity

An important open issue in modern physics is the combination of quantum theory with gravity, creating a unified theory of quantum gravity. This has proven problematic, as the two theories seem incompatible. We believe that the interpretation of quantum physics we choose and that we take it literally may be important in the process of finding a solution to this problem.

One of the reasons finding a theory of quantum gravity proves so hard is that the matter has very limited empirical implications, making it very difficult to collect empirical data on the subject. This means that an empirical generalization from already collected data is impossible. We do, however, have quantum theory and the theory of general relativity, which are both empirically very successful. If we take their theoretical structures literally, and not merely as predictive instruments, we can try to combine them in various ways. Combined theories created this way may yield specific ways to test them empirically. By proceeding this way, we don't have to collect large amounts of data before being able to formulate a first promising hypothesis. Instead, we can collect specific data to test a specific hypothesis.

As mentioned before, combining quantum theory and gravitational theory turns out to be not so easy. This may be due to physicists assuming wrong theoretical structures. Maybe certain theoretical structures are simply impossible (or at least very hard) to combine. Choosing a different theoretical structure may yield different results. Penrose (2004, p. 817) writes on the subject:

> 'The usual perspective, with regard to the proposed marriage between these theories, is that one of them, namely general relativity, must submit itself to the will of the other. There appears to be the common view that the rules of quantum field theory are immutable, and it is Einstein's theory that must bend itself appropriately to fit into the standard quantum mould. Few would suggest that the quantum rules must themselves admit to modification, in order to ensure an appropriately harmonious marriage. [...] Yet, I would claim that there is observational evidence that Nature's view of this union is very different from this!'

So, according to Penrose, quantum mechanics has to be revised in order to make it compatible with gravity. Another author who suggests that a revisiting of the foundations of quantum physics might be due is Flori (2013, p. 2 ff.). Such a revision may also solve any conceptual problems which might exist in relation to the measurement problem. Flori (2013, p. 2) notes:

> 'For the supporters of the claim that indeed there are conceptual problems present in quantum theory, it seems reasonable to try solving fundamental issues before attempting to define a quantum theory of gravity.'

Flori regards *Topos Quantum Theory* as a promising new formalism for quantum mechanics in light of the challenge of a theory of quantum gravity. Penrose, on the other hand, counts on his objective collapse interpretation. We believe that as long as the matter is not resolved empirically it is good that there are a variety of competing research paradigms. This increases the chance of one of them turning out to be right. Hopefully even more approaches will develop in the near future.

14. Conclusion

Quantum physics has thrown doubt on some principles, which we didn't necessarily doubt before. While determinism and value definiteness are seen by some to be necessary parts of a physical theory, we have argued that this is not the case. These principles are neither necessary for gaining scientific knowledge, nor should we expect the world to accord with them.

Support for one of the many interpretations of quantum physics should therefore not be based on these principles. Both conformance and non-conformance with them is acceptable and can thus not serve as a justification for choice of one theory over its alternatives. We have also argued that (except for operational approaches) all interpretations of quantum mechanics are more or less compatible with all discussed ontological positions. So ontology is no decisive ground for theory choice either. Instead, choice of interpretation should mainly be based on the position one takes on the measurement problem – even if the position is that there isn't actually any real problem.

It was not the aim of this work to resolve the measurement problem. We will, however, say that it seems odd (if not impossible) for a deterministic equation like the Schrödinger equation to give rise to non-deterministic phenomena unless there are some non-deterministic defiances of the Schrödinger equation. This suggests some sort of indeterministic mechanism, such as a collapse of the wave function.

We have argued that scientific theories should be *taken literally*, i.e. contradictory theoretical structures should be recognised as significantly distinct, even if they yield identical empirical consequences. They correspond to different research programs and may yield different results when developed further or combined with other theories. In the case of quantum mechanics, this may be relevant for the development of a theory of quantum gravity. Success in the field may depend on a restructuring of quantum mechanics and in particular resolution of any conceptual problems that may remain.

References

[Bell 1964] J. Bell *On the Einstein Podolsky Rosen Paradox*, Physics 1/3, pp. 195–200: 1964.

[Berlin and Kay 1999] B. Berlin and P. Kay *Basic Colour Terms: Their Universality and Evolution*, Stanford: CSLI Publications 1969/1999.

[Boyd 1984] R. Boyd *The Current Status of Scientific Realism*, in: J. Leplin, *Scientific Realism*, pp. 41–821, Berkeley: University of California Press 1984.

[Carnap 1979] R. Carnap *Der logische Aufbau der Welt*, Frankfurt/M: Ullstein 1918/1974.

[Einstein et al. 1935] A. Einstein, B. Podolsky and N. Rosen *Can Quantum-Mechanical Description of Physical Reality be Considered Complete?*, Phys. Rev. 47, pp. 777–780: 1935.

[Englert 2013] B. Englert *On Quantum Theory*: arXiv 2013.

[Everett 1957] H. Everett *Relative State Formulation of Quantum Mechanics*, Review of Modern Physics (29), pp. 454–462: 1957.

[Fließbach 2008a] T. Fließbach *Elektrodynamik, 5th Edition*, Heidelberg: Spektrum Akademischer Verlag 2008.

[Fließbach 2008b] T. Fließbach *Quantenmechanik, 5th Edition*, Heidelberg: Spektrum Akademischer Verlag 2008.

[Fließbach 2009] T. Fließbach *Mechanik, 6th Edition*, Heidelberg: Spektrum Akademischer Verlag 2009.

[Fließbach 2010] T. Fließbach *Statistische Physik, 5th Edition*, Heidelberg: Spektrum Akademischer Verlag 2010.

[Flori 2013] C. Flori *A First Course in Topos Quantum Theory*, Lecture Notes in Physics 868, Springer- Verlag: Berlin/Heidelberg 2013.

[Freedman et al. 1972] S.J. Freedman and J.F. Clauser *Experimental test of local hidden-variable theories*, Phys. Rev. Lett. 28 (938): 1972.

[Garnham and Oakhill 1994] A. Garnham and J. Oakhill *Thinking and Reasoning*, Oxford: Basil Blackwell 1994.

[Goldstein 2013] S. Goldstein *Bohmian Mechanics*, Stanford Encyclopedia of Philosophy: 2013.

[Graemer 2012] D. Graemer *Basissätze*, unpublished: 2012.

[Halliday et al. 2008] D. Halliday, R. Resnick and J. Walker *Fundamentals of Physics, 8th Edition*, Wiley 2008.

[Held 2013] C. Held *The Kochen-Specker Theorem*, Stanford Encyclopedia of Philosophy: 2013.

[Hylsop 2014] A. Hylsop *Other Minds*, Stanford Encyclopedia of Philosophy: 2014.

[Ismael 2009] J. Ismael *Quantum Mechanics*, Stanford Encyclopedia of Philosophy: 2009.

[Jackson 2006] J. D. Jackson *Klassische Elektrodynamik, 4th Edition*, Berlin: Walter de Gruyter 2006.

[Kirkham 1992] R. L. Kirkham *Theories of Truth*, Cambridge: MIT Press 1992.

[Kochen et al. 1967] S. Kochen and E.P. Specker *The Problem of Hidden Variables in Quantum Mechanics*, Journal of Mathematics and Mechanics 17, pp. 59–87: 1967.

[Laudan 1996] L. Laudan *A Confutation of Convergent Realism*, Philosophy of Science 48, pp. 19–49: 1996.

[Lenneberg and Roberts 1953] E. H. Lenneberg and J. M. Roberts *The denotata of color terms*, Paper presented at the Linguistic Society of America, Blookington, Indiana: 1953.

[Norton 2003] J. D. Norton *Causation as Folk Science*, Philosopher's Imprint 3/4: 2003.

[Penrose 2004] R. Penrose *The Road to Reality: A Complete Guide to the Laws of the Universe*, New York: Vintage Books 2004.

[Popper 2005] K. Popper and H. Keuth (Publ.) *Logik der Forschung, 11th Edition*, Tübingen: Mohr Siebeck 1935/2005.

[Putnam 1975] H. Putnam *What is Mathematical Truth?*, in: E. Nagel, P. Suppes and A. Tarski (eds.), *Logic, Methodology and Philosophy of Science*, pp. 240–251, Stanford: Stanford University Press 1975.

[Schlick 1918] M. Schlick *Allgemeine Erkenntnislehre*, Berlin: Verlag von Julius Springer 1918.

[Schrödinger 1935] E. Schrödinger *Die Gegenwärtige Situation in der Quantenmechanik*, Naturwissenschaften 23 (49), pp. 807–812: 1935.

[Schurz 1995] G. Schurz *Skriptum: Vorlesung Erkenntnistheorie*, unpublished: 1995.

[Schurz 2006] G. Schurz *Skriptum: Grundprobleme der Wissenschaftstheorie*, Fernuniversität Hagen: 2006.

[Schurz 2009] G. Schurz *When Empirical Success Implies Theoretical Reference: A Structural Correspondence Theorem*, British Journal for the Philosophy of Science 60/1, pp. 101–133.

[Schurz 2014] G. Schurz *Einführung in die Wissenschaftstheorie, 4th Edition*, Darmstadt: Wissenschaftliche Buchgesellschaft 2006/2014.

[Schurz 2013] G. Schurz *Philosophy of Science: A unified approach*, New York: Routledge 2013.

[Spekkens 2011] R. Spekkens *Foundations of Quantum Mechanics*, Lecture given at the Perimeter Institute for theoretical Physics, Ontario: 2011.

[Swoyer 2003] C. Swoyer *The Linguistic Relativity Hypothesis*, Stanford Encyclopedia of Philosophy: 2003.

[Uebel 2011] T. Uebel *Vienne Circle*, Stanford Encyclopedia of Philosophy: 2011.

[Vaidman 2014] L. Vaidman *Many-Worlds Interpretation of Quantum Mechanics*, Stanford Encyclopedia of Philosophy: 2014.

[Van Fraasen 1980] B. Van Fraasen *The Scientific Image*, Oxford: Ckaredon Press 1980.

[Van Fraasen 1991] B. Van Fraasen *Quantum Mechanics: An Empiricist View*, Oxford: Oxford University Press 1991.

[Verein Ernst Mach 1981] Verein Ernst Mach *Wissenschaftliche Weltauffassung – Der Wiener Kreis*, reprinted in: H. Rudolf (ed.) and Rutte, Heiner (ed.): *Otto Neurath – Gesammelte philosophische und methodologische Schriften, Band 1*, Wien: Verlag Hölder-Pichler-Temspky 1929/1981.

[Weingartner 1978] P. Weingartner *Wissenschaftstheorie I: Einführung in die Hauptprobleme, 2nd Edition*, Stuttgart: frommann-holzboog 1978.

[Worrall 1989] J. Worrall *Structural Realism: The Best of Both Worlds?*, Dialectica 43/1–2, pp. 99–124: 1989.

[Young 1802] T. Young *The Bakerian Lecture: On the Theory of Light and Colours*, Philosophical Transactions of the Royal Society of London 82, pp. 12–48: 1802.

List of Figures

Figure 1: Diffraction of wave at a single slit. Image used with the permission of Philip Graemer. 51

Figure 2: Illustration of the principle of superposition showing constructive interference (top) and destructive interference (bottom). Image used with the permission of Philip Graemer. 52

Figure 3: Scattering of a wave at a double slit showing the resulting interference pattern. Image used with the permission of Philip Graemer. 53

Part 2: Quantum Logic

Introduction

This part of the book examines approaches to quantum logic which consider classical logic to be defeated by the findings of quantum mechanics and therefore propose a new logical system to deal with quantum mechanical phenomena or even replace classical logic. Classical logic is based on two fundamental principles: (a) the *principle of bivalence*, according to which one of the two truth values (*true* and *false*) can be assigned to every proposition – independent of our ability to know which of the values a certain proposition has, and (b) the *principle of extensionality*, according to which the truth value of compound sentences is determined by the truth values of its components. The findings in the area of quantum mechanics, namely *Heisenberg's uncertainty principle*, do not allow for a sentence like "The electron e can be found at position x and has a momentum p." to be assigned the truth value *true*. When the position of a particle has been determined, an equally exact determination of the momentum is impossible. Further, quantum objects are subject to *wave-particle duality*, meaning they can act like waves and particles under certain circumstances. Heisenberg's uncertainty principle and wave-particle duality contradict the principles of bivalence and extensionality according to some of the quantum logicians' argumentation. It is argued that this is due to the impossibility of determining the truth value of a compound sentence about quantum mechanical phenomena from the truth values of its components and to the impossibility of assigning truth values to certain statements about quantum objects in general.

This problem will be addressed in the following. In chapter 1, the fundamentals of quantum mechanics and classical logic relevant to the discussion will be described. The failure of the *distributive law* is a main argument for the invalidity of classical logic when it comes to statements about quantum mechanical phenomena. It will be discussed in chapter 2. Arguments by John von Neumann and Garrett Birkhoff (1936) and Hilary Putnam (1968) and counter-arguments by Karl Popper (1968), Gerhard Schurz (2014), Michael Dummett (1976) and John Stachel (1986) will be reconstructed and discussed. As a conclusion it is found that by correct classical assignment of truth values to the statements in question and by interpreting the sentence connectives involved classically the distributive law does not fail to hold for the quantum mechanical phenomena discussed. Further it will be shown that classical logic cannot be defeated in case the underlying structure is no longer isomorphic to a Boolean algebra, because a comparison of the relevant properties of the structures is then no longer possible.

Chapter 3 is a closer examination of Hilary Putnam's paper "Is logic empirical?" (1968). It comprises his invariance argument, according to which the meaning of the logical connectives does not change when the distributive law is omitted from classical logic, and his argument for bivalence in his quantum logical system, which he justifies with a certain interpretation of statements about quantum objects. The counter-arguments by John Bell and Michael Hallett from their paper "Logic, Quantum Logic and Empiricism" (1982) will be reconstructed, leading to the conclusion that neither can invariance be claimed to hold for Putnam's system nor is it compatible with bivalence.

The results are summarised and discussed in chapter 4. In a nutshell, this text comes to the conclusion that classical logic cannot be defeated by the findings of quantum mechanics as the discussed approaches propose, since they are based on a false or non-classical assignment of propositions or on a change of meaning of the logical connectives and therefore should not have been considered to be arguments against classical logic in the first place.

A fundamental assumption underlying the examination of the relationship between classical logic and quantum mechanics is that logic can be empirically founded, or at least motivated. Otherwise, the question whether there is a need for a new logic to describe quantum mechanical phenomena would have to be negated right away. Since the presented approaches necessarily make this assumption, it will not be questioned in this text.

1. Fundamentals of Classical Logic and Quantum Mechanics

This chapter presents the logical and physical fundamentals for understanding the following philosophical analysis of quantum logic. The phenomena of quantum mechanics in the approaches to quantum logic discussed in the chapters to come, namely wave-particle duality and Heisenberg's uncertainty principle, are illustrated by means of the double-slit experiment in section 1.1. The fundamental principles of classical logic are summarised in section 1.2 with focus on the principles of extensionality and bivalence, to which the quantum-logical approaches in this paper will be contrasted.

1.1 Peculiarities of Quantum Mechanics

Classical mechanics is a deterministic theory. This means that each property of a system has a specified value at each point in time, which makes future states of a system computable when the initial state is known[13]. Wave properties and particle properties are mutually exclusive in classical physics since particles are described as "localized bundles of energy and momentum"[14] and "[a] wave, in contrast, is a disturbance spread over space. It is described by a wave function $\psi(r,t)$ which characterizes the disturbance at the point r at time t."[15]

These classical definitions cannot be applied to quantum mechanical objects since they show properties of particles *and* waves, which led to the assignment of a wave function $\psi(x,t)$ to classical particles *and* waves, the squared absolute $|\psi(x,t)|^2$ of which "gives the probability of finding it at a point x at time t"[16]. This property of quantum objects is called wave-particle

---

13 Cf. Auletta, G., Fortunato, M., Parisi, G. (2009): *Quantum Mechanics*, Cambridge University Press, 7.
14 Shankar, R. (1994): *Principles of Quantum Mechanics* (2nd edition), Plenum Press, New York, 107.
15 Ibid. [Emphasis in original].
16 Ibid., 113.

duality. It can be seen in the double-slit experiment in which single particles are registered on an observation screen after passing two slits (figure 1).

Figure 1: Apparatus of the double-slit experiment.[17]

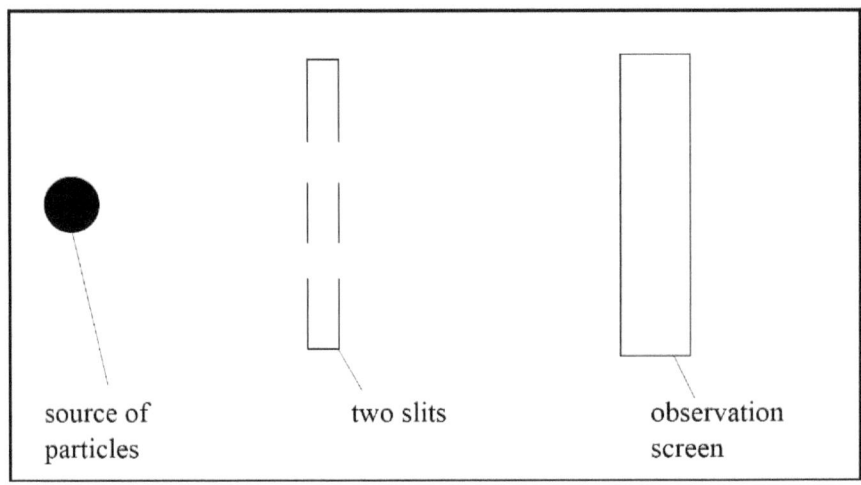

When the experiment is carried out with classical particles, like munition, the observation screen would show a pattern with maxima directly behind the slits, see figure 2(a). Classical waves, e.g. water waves, would be fractured on the slits and new circular waves would spread from there, leading to an interference pattern on the observation screen as a result of constructive and destructive interference, see figure 2(b).[18] The maxima (minima) are at those spots at which the distance of the observation screen to one of the slits is an integer of the wavelength (half of the wavelength) bigger or smaller than the distance from the observation screen to the other slit[19]. If the experiment is carried out with electrons as quantum objects, an interference pattern is detected on the observation screen (wave property)

---

17 Cf. Feynman, R. (2010): *The Feynman Lectures on Physics*, Volume 3: Quantum Mechanics, Basic Books, 1–2.
18 Cf. Ibid., 1.2.
19 Cf. Ibid., 1–3.

while at the same time they are registered individually (particle property), see figure 2(c)[20].

Figure 2: Patterns observed in the double-slit experiments (a) with classical particles, (b) classical waves and (c) quantum objects.

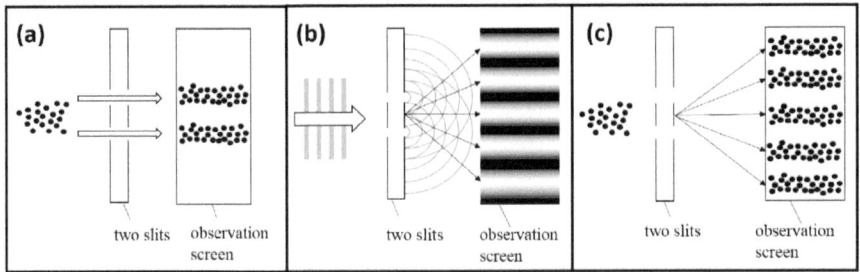

Given that particles have a sharp position and electrons as quantum objects are not dividable it should be possible to determine through which of the two slits each single particle passes (the which-way information). But the pattern on the observation screen changes drastically as soon as a detector is placed behind the slits to determine the which-way information: there is no interference pattern anymore, but a pattern as is observed when carrying out the experiment with particles as displayed in figure 2(a). So it is not possible to know the way of the electron without changing the results of the experiment.[21] This has been demonstrated for all quantum objects[22].

Summing up, the following relevant phenomena can be seen from the experiment:

<u>Wave-particle duality</u>: Quantum objects show properties of waves and particles. In the double-slit experiment without detectors they are registered individually on the observation screen and build up an interference pattern as a wave characteristic.

<u>Heisenberg's uncertainty principle</u> (HU): Determining precisely the position x of a quantum object leads to a more imprecise determinability of its momentum p, which is described by the formula $\Delta x \Delta p \geq \frac{\hbar}{2}$. (Electrons

---

20 Cf. Ibid., 1–5.
21 Cf. Feynman (2010), 1–7.
22 Cf. Straumann, N. (2013): *Quantenmechanik: Ein Grundkurs über nichtrelativistische Quantentheorie*, 2. Auflage, Springer Spektrum, 23.

and other material particles do have a wavelength λ which is related to the momentum by the de-Broglie-relation $λ = h/p$[23].) The formula means that the certainty of position ($Δx$) multiplied by the certainty of momentum ($Δp$) cannot be determined more precisely than the reduced Planck constant ($ℏ$). In the experiment, this can be seen from the interference pattern being smeared out as soon as the way through the double-slit is determined using a detector[24].

Moreover, it is impossible to determine the which-way information for one single electron beforehand. Only statistical predictions are possible with regard to the pattern detected on the observation screen and the behaviour of electrons.

1.2 Classical Logic

By classical logic we mean classical propositional logic. Here, as in many logical systems, formalisation is used to be able to make general statements about the truth and falsity of sentences and the validity of inferences and inference schemes. For this purpose, small letters $p, q, r, …$ stand for propositions (like "It's raining.") and truth-functional operators ($∨, ∧, →, ↔, ¬$) stand for logical connectives (like the and-connection in "It's raining and the street is wet.").

Formalisation enables us to study not only particular sentences and arguments but the truth/falsity of sentence forms and the validity/invalidity of inference schemes. The definitions of logically true (L-true) sentences and logically valid (valid) inferences are:

<u>Logically true:</u> A sentence is logically true (for short: L-true) iff its (logical) sentence form is true for every possible interpretation of its non-logical symbols. We then also call this sentence form logically true.[25]

<u>Logically valid:</u> An inference is logically valid (for short: valid) iff for the logical form of this inference the following holds: every possible interpretation which makes all premises true, also makes the conclusion true. – Equivalently expressed: … iff there is no interpretation which makes all the premises true, but

---

23 Cf. Shankar, R. (1994), 112.
24 Cf. Feynman (2010), 1–11.
25 Schurz, G. (2013b): *Philosophy of Science, A Unified Approach*, Routledge, New York, 110.

the conclusion false. – If an inference is valid, then we say that its conclusion follows (logically) from its premises.[26]

The logical connectives have unique truth conditions which are summarised in table 1. For any connection of sentences with sentence connectives it is possible to decide whether it is true or false using truth conditions. The sentence "It's not raining." ($\neg p$) for example is true if and only if (iff) the sentence "It's raining." is false.

Table 1: Connectives in classical logic with truth conditions[27]

Connective	Name	Truth conditions
$\neg p$	Negation	$\neg p$ is true iff p is false, otherwise false.
$p \wedge q$	Conjunction	$p \wedge q$ is true iff p is true and q is true, otherwise false.
$p \vee q$	Disjunction	$p \vee q$ is false iff p is false and q is false, otherwise true.
$p \to q$	Implication	$p \to q$ is false iff p is true and q is false, otherwise true.
$p \leftrightarrow q$	Biconditional	$p \leftrightarrow q$ is true iff p and q have the same truth value.

L-true (tautological) and L-false (contradictive) sentences (symbols: $\top$ resp. $\bot$) have a special status because they are true (resp. false) just because of their logical form: the former are true for each truth value assignment of the propositions they consist of, like $p \vee \neg p$ (Law of the Excluded Middle), the latter are false for each truth value assignment of their propositions, like $p \wedge \neg p$. Contingent sentences are sentences for which the truth value depends on the truth value of their components in the sense of varying with varying component truth values, the simplest example being p which is true if the sentence formalised as p is true and false if the underlying sentence is false. The truth values of the sentence connectives for each possible assignment of truth values to the sentences p and q, as well as the terms L-true and L-false are illustrated in a truth table in table 2.

---

26 Ibid., 112.
27 Cf. Schurz, G.(2013a): *Logik I: Einführung in die Aussagen- und Prädikatenlogik*. Available at: https://www.phil-fak.uni-duesseldorf.de/fileadmin/Redaktion/Institute/Philosophie/Theoretische_Philosophie/Schurz/scripts/SkriptVLLogikI-Neu.pdf, 25–26. Accessed: 20.02.2015.

Table 2: Truth values of sentence connectives and logically true and false sentences for all possible truth values of the propositions p and q (t=true and f=false)[28]

Truth-values of propositions		Negation	Conjunction	Disjunction	Implication	L-true sentence	L-false sentence
p	q	$\neg p$	$p \wedge q$	$p \vee q$	$p \rightarrow q$	$p \vee \neg p$	$p \wedge \neg p$
t	t	f	t	t	t	t	f
t	f	f	f	t	f	t	f
f	t	t	f	t	t	t	f
f	f	t	f	f	t	t	f

Classical logic is isomorphic to a Boolean algebra, which is defined as follows:

> "A Boolean algebra is a system consisting of a set S and two operations, ∩ and ∪ (cap and cup), subject to the following axioms. For all sets a, b, c, that are members of S:
> 1. $a \cap (b \cap c) = (a \cap b) \cap c$
>   Also $a \cup (b \cup c) = (a \cup b) \cup c$ (associativity)
> 2. $a \cap b = b \cap a$
>   Also $a \cup b = b \cup a$ (commutativity)
> 3. $a \cap (b \cup c) = (a \cap b) \cup (a \cap c)$
>   Also $a \cup (b \cap c) = (a \cup b) \cap (a \cup c)$ (distributivity)
> 4. There belong to S two elements, 0 and 1, with the properties $a \cup 0 = a$; $a \cap 1 = a$ (identity)
> 5. For each set a in S there exists a set a' with the properties that $a \cup a' = 1$; $a \cap a' = 0$. (complementation)."[29]

Classical logic can be represented as a Boolean algebra by the assignments shown in table 3.

---

28 Cf. Ibid., 25–41.
29 "Boolean Algebra", *The Oxford Dictionary of Philosophy*, Second Edition revised (2008), 45.

Table 3: Isomorphism between Boolean algebras and classical logic[30]

Objects in Boolean algebras	Objects in Classical logic
Union	Disjunction
Intersection	Conjunction
Complementation	Negation
Identity	Equivalence classes of L-true (⊤) and L-false (⊥) sentences.
Set members	Sets of logically equivalent sentences (equivalence classes).

Two semantic principles are essential to classical logic (my own translation): Principle of extensionality: "A sentence connective is truth-functional resp. extensional, if the truth value of its application is always and uniquely determined by the truth values of its arguments."[31]

Principle of bivalence: "*Any statement is either true or false (principle of bivalence).*

This *principle of bivalence* can be split into two parts:

o Any statement is either true or false, there is no 'third' truth value (*Law of Excluded Middle*, formalised: A or not-A).
o No statement can be true and false at the same time (*Law of Noncontradiction*, formalised: not(A and not-A)."[32]

The former (section 1.1) described impossibility of determining a quantum object's position and momentum with a certain precision due to HU[33] is in certain interpretations a defeater for the principle of extensionality. For example, the statements p and q can be true simultaneously, without their conjunction being true for small intervals $[a, b]$ and $[c, d]$:

---

30 Cf. Schurz, G. (2014): *Skript Logik II*, available at: https://www.phil-fak.uni-duesseldorf.de/fileadmin/Redaktion/Institute/Philosophie/Theoretische_Philosophie/Schurz/teaching_materials/VLLogikII2014.doc, 67. Accessed: 16.02.2015.
31 Schurz, G. (2013a), 12.
32 Ibid., 16. [Emphasis in original]
33 It is always possible to calculate probabilities for position and momentum though.

p = "The particle's position is in the interval [*a*, *b*]."
q = "The particle's momentum is in the interval [*c*, *d*]."

According to the principle of extensionality, the truth value of a compound sentence can be determined from the truth value of its components and a conjunction is true iff both of its components are true, which is not automatically the case here since the conjunction $p \wedge q$ might violate HU.

A similar point can be made with regard to wave-particle duality. Analysed separately the statements *r* and *s* seem to have a different meaning than there disjunction has:

r = "The particle passes through the left slit."
s = "The particle passes through the right slit."

In case the which-way information is not determined and accordingly the disjunction $r \vee s$ is interpreted as if there was no detector in the double-slit experiment, *r* and *s* seem to be false when analysed separately, whereas their disjunction $r \vee s$ seems to be true. This contradicts classical logic because for a disjunction to be true at least one of its disjuncts has to be true. *r* resp. *s* can only be true in case the which-way information has been determined. Therefore, the statements and their disjunction are subject to different experimental setups in this interpretation.

These problems were addressed by John von Neumann and Garrett Birkhoff as well as Hilary Putnam and shall be discussed in the following.

2. Quantum logic and the distributive law

An argument for the refusal or restriction of the validity of classical logic when it comes to quantum mechanics is that the distributive law

$$p \wedge (q \vee r) \leftrightarrow (p \wedge q) \vee (p \wedge r)$$

is not valid for certain quantum mechanical statements. In table 4 it is demonstrated that it is valid in classical logic since the left and the right side of the argument have the same truth value for any truth value assigned to the propositions p, q, r – both sides are either true or false and therefore the truth condition of the biconditional connective is satisfied.

Table 4: Equivalence of the left and the right side of the distributive law shown in a truth table

Truth-values of propositions			Distributive law		
p	q	r	$p \wedge (q \vee r)$	$\leftrightarrow$	$(p \wedge q) \vee (p \wedge r)$
t	t	t	t	t	t
t	t	f	t	t	t
t	f	t	t	t	t
f	t	f	f	t	f
f	t	t	f	t	f
f	f	f	f	t	f

Two types of arguments for its failure in quantum mechanics are discussed in the following:

1) Birkhoff and von Neumann construct a counter-example in which $p \wedge (q \vee r)$ is tautologically true, while the conjunctions $(p \wedge q)$ and $(p \wedge r)$ are false.
2) Putnam proposes a logical structure in which the left and the right side of the distributive law have different meanings for mathematical reasons.

2.1 Birkhoff and von Neumann's argument

In their paper "The Logic of Quantum Mechanics" (1936), Garrett Birkhoff and John von Neumann argue for the failure of the distributive law for quantum mechanical propositions. They introduce a set of propositions in which the left side and the right side are not equivalent:

> "That it [the distributive law, AS] *does* break down is shown by the fact that if a denotes the experimental observation of a wave-packet ψ on one side of a plane in ordinary space, a' correspondingly the observation of ψ on the other side, and b the observation of ψ in a state symmetric about the plane, then (as one can readily check):
>
> $b \cap (a \cup a') = b \cap 1 = b > 0 = (b \cap a) = (b \cap a') = (b \cap a) \cup (b \cap a')$"[34]

In logical symbols their argument is represented as

$$b \wedge (a \vee \neg a) \leftrightarrow b \wedge \top \leftrightarrow b \not\leftrightarrow \bot \leftrightarrow (b \wedge a) \leftrightarrow (b \wedge \neg a) \leftrightarrow (b \wedge a) \vee (b \wedge \neg a).$$

The argument is based on the assumption that the observation of a particle ψ on the one *and* the other side of a plane at the same time is a tautology ($a \vee \neg a$), while observing the one resp. the other side at the same time with a symmetric observation of the plane is regarded a contradiction ($b \wedge a$ resp. $b \wedge \neg a$). The reason for this assumption can be considered to be wave-particle duality, due to which the possibility of determining the position (particle property) and at the same time observing the whole plane (which will without disturbance result in an interference pattern characteristic for waves) is excluded.

To illustrate their argument more clearly, it shall be discussed applied to a double-slit experiment performed with one single electron. The set of the propositions a, $\neg a$ and b is in this case

- a = "Detection of the electron at the left slit."
- $\neg a$ = "Detection of the electron at the right slit."
- b = "Observation of the double-slit without determining the which-way information."

---

[34] Birkhoff, G., von Neumann, J. (1936): "The Logic of Quantum Mechanics", *Annals of Mathematics* 37, 823–843, 831. [Emphasis in original, notation aligned to the rest of this text.]

and is illustrated in figure 3, where the path of the electron is indicated by an arrow.

Figure 3: Propositions in Birkhoff and von Neumann's argument against the distributive law applied to the double-slit experiment (D=detector).[35]

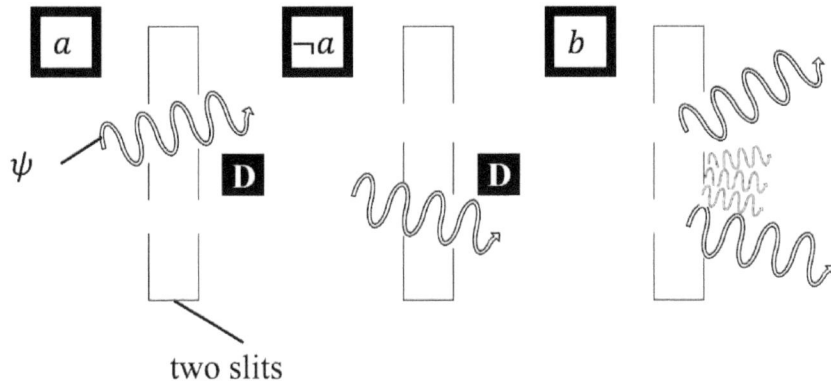

two slits

The two possible paths that the electron can take through the double-slit are denoted by the propositions a and $\neg a$. In reality, the detection of the electron at one of the slits (left or right) will result in a particle pattern on the observation screen when enough electrons have been sent through the apparatus. Proposition b denotes the observation of the double-slit without detectors which would result in an interference pattern on the observation screen for a sufficient amount of electrons. It is indicated by several arrows that the which-way information is not available is proposition b is true.

Loosely formalised, Birkhoff and von Neumann's argument is:
<u>Left side (from left to right)</u>: Observing the double-slit without which-way information AND (Detection of the electron at the left slit OR detection of the electron at the right slit) is equal to observing the double-slit without which-way information AND a tautology, which is equal to observing the double-slit without which-way information.
<u>Right side (from right to left)</u>: (Observing the double-slit without which-way information AND detection of the electron at the left slit) OR (Observing the double-slit without which-way information AND detection of the

---

35 This figure is based on a sketch by Gerhard Schurz [P.C., 12/2014].

electron at the right slit) is equal to (Observing the double-slit without which-way information AND detection of the electron at the left slit), which is equal to (Observing the double-slit without which-way information AND detection of the electron at the right slit) and both are contradictions.

So it is assumed on the one hand that the detection of the electron at the left and the right slit at the same time ($a \vee \neg a$) is a tautology, whereas the observation of the double-slit as a whole and the detection of the electron at one of the two slits at the same time is interpreted as a contradiction. In this example this is due to the interference pattern smearing out as soon as the which-way information is determined, which makes the truth of a sentence like "The particle passes the left slit and acts like a wave." impossible.

These assumptions can ad hoc be criticised as follows. As illustrated in figure 3, detection of the electron at the left or right slit is not the same as an observation without detection, since determination of the which-way information will always result in the electron behaving as a particle, which is indicated by one single arrow in propositions a and $\neg a$, whereas the experiment without detectors will result in the electron behaving like a wave which cannot be assigned a sharp position, indicated by many arrows in proposition b. Karl Popper's criticism of this argument will be discussed in the following.

2.1.1 Popper's criticism

Karl Popper criticises the set of propositions which Birkhoff and von Neumann chose in their argument in that it would not be in accordance with classical logic which made their argument invalid:

> "Nothing in this argument is specific to quantum mechanics: indeed, we may substitute an elephant in 'ordinary observation space' for the wave-packet. For if a denotes the experimental observation or position of anything, be it a wave-packet or an elephant (even a classical mass-point may be included, though it comes nearest to being dubious) on one side of a plane in ordinary observation space, then a' (the complement of a which according to Birkhoff and von Neumann is the ordinary classical set-theoretic orthocomplement) does not 'correspondingly' denote its observation or position on the other side. a' denotes, rather, the property 'not on the one side'. For wave-packets, elephants, or even classical mass-points, this is perfectly compatible with the property denoted earlier by b; that is to say with the property 'symmetric about the plane'. It is also compatible with many

other properties. Thus $b = b \wedge a'$ and $b \neq 0 = b \wedge a \neq b \wedge a'$ and so the thought experiment of Birkhoff and von Neumann breaks down."[36]

Thus, according to Popper, if a denotes observation of the one side, $\neg a$ does not denote observation of the other side, but in accordance with the classical definition of the negation connective an observation made *not on the one side*, which is a proper subset of proposition b and thus b can be substituted for $b \wedge \neg a$. As a consequence, the truth values of $b \wedge a$ and $b \wedge \neg a$ are not identical, as Birkhoff and von Neumann argued, but distinct, the former as a contradiction being false in any case, the latter being contingent as b can be substituted for it and which is therefore true if b is true.

Applied to the example of the double-slit experiment introduced above to clarify Birkhoff and von Neumann's argument, $\neg a$ would no longer stand for the detection on the right side, but for

$\neg_2 a$ = "Observation with no detection at the left slit".

Thus, in the following the negation used by Birkhoff and von Neumann will be denoted $\neg_1$ and Popper's negation $\neg_2$. Loosely formalised Popper's argument is then:

<u>Left side</u> ($b \leftrightarrow b \wedge \neg_2 a$): Observing the double-slit without which-way information is equal to observing the double-slit without which-way information AND observation without detection at the left slit.

<u>Right side</u> ($b \not\leftrightarrow 0 \leftrightarrow b \wedge a \not\leftrightarrow b \wedge \neg_2 a$): Observing the double-slit without which-way information is not equal to the contradiction "Observing the double-slit without which-way information AND detection of the electron at the left slit" which in turn is not equal to observing the double-slit without which-way information AND observation without detection at the left slit.

Popper's argument seems to be more compatible with quantum mechanics, especially with wave-particle duality, because it takes into account the fact that the experiment has different outcomes depending on whether or not a detector is integrated. This is why $b \wedge a$ is a contradiction. Figure 4 demonstrates the argument including the changes proposed by Popper.

---

36 Popper, K. (1968): "Birkhoff and von Neumann's Interpretation of Quantum Mechanics", *Nature* 219, 682–685, 685. [Notation aligned to fit the rest of this text.]

Figure 4: Popper's criticism applied to Birkhoff and von Neumann's argument.

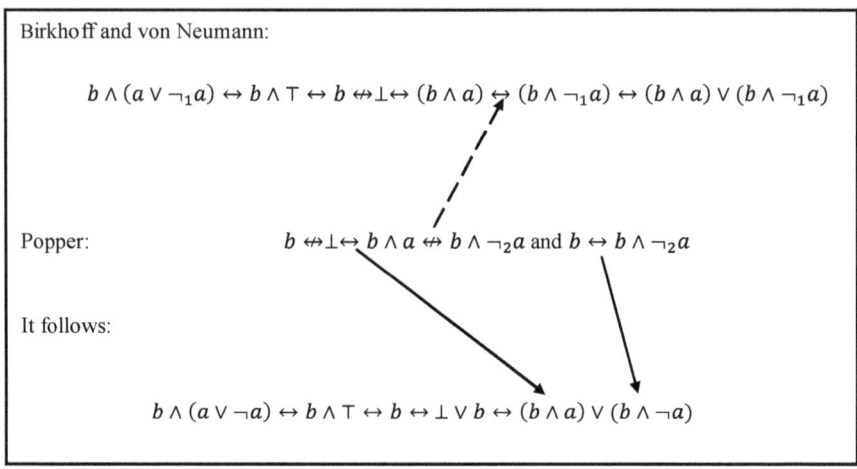

The dashed arrow points to the part of the argument that Popper rejects: detecting the electron at the left slit and at the same time observing the double-slit without which-way information is a contradiction and therefore is not equal to the observation without detection at the left slit while observing without which-way information (which can be regarded identical, as argued). The continuous arrows point at the parts of the argument where Popper's changes have to be applied: $b \wedge a$ is equal to a contradiction and can be replaced therefore by $\bot$ and $b \wedge \neg_2 a$ can be replaced by b. As a result, the left side as well as the right side of the argument are equal to b, the observation of the double-slit without which-way information. Thus, on the grounds of the classical meaning of negation alone, Birkhoff and von Neumann's argument is not valid according to Popper.

2.1.2 Schurz's criticism

A slight variation of the argument was proposed:

> "To illustrate why the distributive law fails, consider a particle moving on a line and let:
> a = 'the particle has momentum in the interval $\left[0, +\frac{1}{6}\right]$'
>
> b = 'the particle is in the interval $[-1, 1]$'
>
> c = 'the particle is in the interval $[-1, 3]$'

We might observe that:

$$a \wedge (b \vee c) = \text{true}$$

In other words, that the particle's momentum is between 0 and +1/6, and its position is between −1 and +3. On the other hand, the propositions a ∧ b and a ∧ c are both false, since they assert tighter restrictions on simultaneous values of position and momentum than is allowed by the uncertainty principle in quantum mechanics. So:

$$(a \wedge b) \vee (a \wedge c) = \text{false}$$

Thus the distributive law fails."[37]

Just as in the argument proposed by Birkhoff and von Neumann, a set of propositions is proposed in which the right side of the distributive law is false due to the findings of quantum mechanics, here because of HU. On the right side of the argument, the position and momentum coordinates are determined with greater certainty than HU allows. On the left side of the argument, the disjunction of *b* and *c* is considered to be equivalent to the union of the intervals described by these propositions, [−1,1] ∪ [1,3] = [−1,3], which is compatible with HU.

Gerhard Schurz[38] criticises the set of propositions in this argument. According to quantum mechanics it is not possible to make a statement about a particle (e.g. an electron) *being* in a defined interval or *having* a defined momentum before measurement, but only about a measurement carried out on the particle to find out its position or momentum leading to a certain result. It is only possible to make statements about the result of a certain measurement of position or of momentum, and of the probability of this result. This is why in quantum mechanics statements about the position or momentum of a particle *before* measurement are always statements about the possible results of a position or momentum measurement. Thus, in accordance with quantum mechanics, the predictions possible before measurement are rather:

---

37 Giedra, Haroldas (2014): "Proof System for Logic of Correlated Knowledge", 31. Available at http://www.mii.lt/files/mii_dis_2014_giedra.pdf . Downloaded on 20.02.2015.
38 This argument is reconstructed here based on personal communication with Prof. Gerhard Schurz [P.C., 12/2014].

a^* = "The result of a measurement of the particle's momentum will be in the interval $\left[0, +\frac{1}{6}\right]$."

b^* = "The result of a measurement of the particle's position will be in the interval $[-1,1]$."

c^* = "The result of a measurement of the particle's position will be in the interval $[1,3]$."

Measuring the particle's position and its momentum at the same time is restricted by HU. Therefore, the statements about the position of the particle, b^* and c^*, are false when the statement about the position, a^*, is true. It is not possible to carry out the measurement corresponding to b^* and c^* as soon as the momentum has been specified. On the other hand, if either b^* or c^* is true, a prediction of the momentum as described by a^* is no longer possible which thus leads to a^* being false. Summing up: The propositions a^* and b^* and a^* and c^* describe mutually exclusive properties because of HU. In figure 5 it is illustrated that the distributive law holds for the aligned set of propositions proposed by Schurz – exemplarily for the case of a^* being true making b^* and c^* false. Since the main sentence connectives on the left ($\wedge$) and on the right ($\vee$) side of the argument are false, the biconditional is true.

Figure 5: *Validity of the distributive law following the argumentation of Schurz, exemplarily for a^* being true, b^* and c^* being false.*

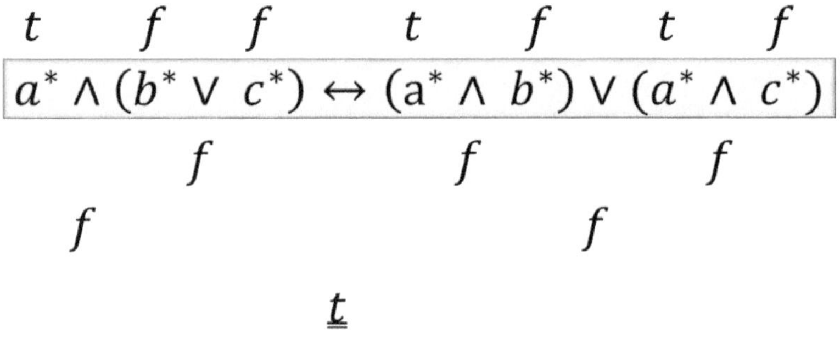

$$\begin{array}{ccccccc} t & f & f & t & f & t & f \\ \hline a^* \wedge (& b^* \vee & c^*) & \leftrightarrow (a^* \wedge & b^*) \vee & (a^* \wedge & c^*) \\ & f & & & f & & f \\ f & & & & & f & \\ & & & t & & & \end{array}$$

Both Popper's and Schurz's arguments account for quantum mechanics more accurately. Schurz's argument reflects HU more accurate than the original argument because it accounts for the fact that simultaneous measurement of position and momentum is in quantum mechanics restricted by HU.

2.2 Putnams's argument

Hilary Putnam defines statements about quantum mechanical system as "statements of the form $m(s) = r$ – 'the magnitude m has the value r in the system s'"[39]. Each of these statements corresponds to a one-dimensional subspace in the vector space which describes the system in question mathematically and the subspaces corresponding to all possible values r_n span the vector space (i. e. they define its dimension – meaning mathematically that they are mutually orthogonal and a set of basis vectors). Figure 6 illustrates this for the three-dimensional case, meaning that the considered magnitude has three possible values. The one-dimensional subspaces (vectors) $S(p)$, $S(q)$ and $S(r)$ each correspond to a statement regarding the value of the magnitude in and span together the vector space $H(s)$.

Figure 6: Three-dimensional example of the assignment of propositions $S(p)$, $S(q)$ and $S(r)$ to vectors in the vector space $H(s)$ describing the system.

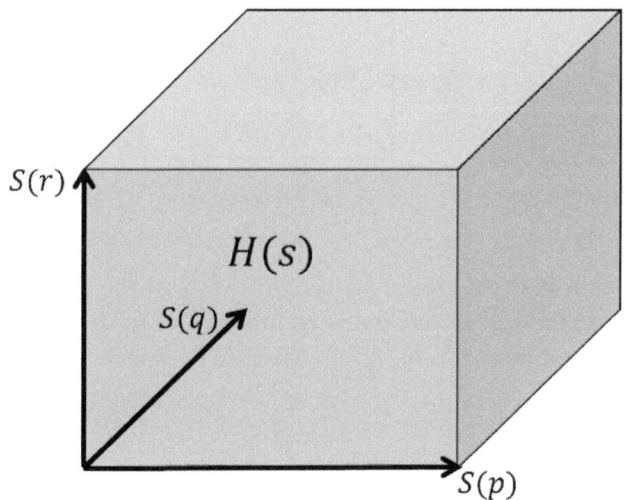

---

39 Putnam, H. (1968): "Is logic empirical?", Boston Studies in the Philosophy of Science 5, Dordrecht, Holland: Reidel, 216–41. Page numbers in the text are taken from the reprint titled "The Logic of Quantum Mechanics" in Putnam (1979): Philosophical Papers Vol. 1: 174–197, 177. [Emphasis in original.]

The sentence connectives are defined as follows in Putnam's system:

"$S(p \vee q)$ = the *span* of the spaces $S(p)$ and $S(q)$, (1)

$S(p \wedge q)$ = the *intersection* of the spaces $S(p)$ and $S(q)$, (2)

$S(\neg p)$ = the orthocomplement of $S(p)$. (3)"[40]

In contrast to Boolean algebras the disjunction of two propositions does not correspond to the union of two sets but is defined as the span of the vectors of which it consists. Negation is defined as orthocomplement, but in space and not for sets. Figure 7 illustrates the mathematical meaning of Putnam's sentence connectives exemplarily for the three-dimensional case.

Figure 7: Mathematical meaning of the sentence connectives as proposed by Putnam for the three-dimensional case.

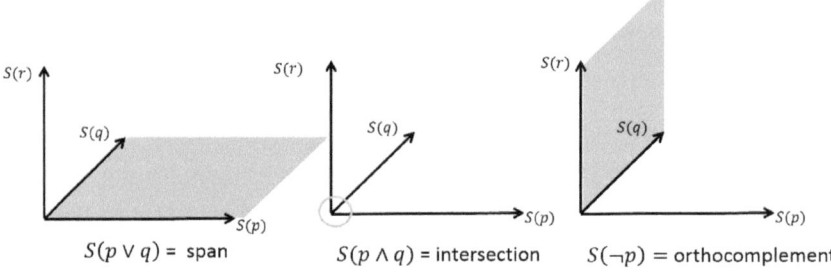

In this system Putnam argues for the invalidity of the distributive law for quantum mechanical propositions as follows. Considering the properties position, S_i, and momentum, T_j, of a particle he states:

"[...] $S_1 \vee S_2 \vee \ldots \vee S_R$ (1)

is a true state description in quantum logic and so is

$T_1 \vee T_2 \vee \ldots \vee T_R$ (2)

In words: Some S_i is a true state description (1')

and   Some T_j is a true state description. (2')

[...] However [...] the conjunction $S_i \wedge T_j$ is inconsistent for all i, j."[41]

---

40 Ibid., 178. [Notation aligned to fit the rest of this text.]
41 Ibid., 184–185. [Notation aligned to fit the rest of this text.]

Thus, he evaluates the disjunction over all possible position resp. momentum coordinates to be true, while he evaluates the conjunction of two specific values, $S_i \wedge T_j$, to be false for any possible pairing (for all i, j). Assuming that position S has been specified by measurement and the possible momentum coordinates are $T_1 \vee T_2 \vee \ldots \vee T_R$, he reasons

$$S \wedge (T_1 \vee T_2 \vee \ldots \vee T_R) \leftrightarrow S.^{42}$$

In words: The intersection of the position (S) and the whole vector space ($T_1 \vee T_2 \vee \ldots \vee T_R$ spans the vector space since this disjunction represents all possible values of the considered magnitude – see figure 8, left) is equal to the position.

On the other hand, he argues that

$$(S \wedge T_1) \vee (S \wedge T_2) \vee \ldots \vee (S \wedge T_R) \leftrightarrow \bot \vee \bot \vee \ldots \vee \bot \leftrightarrow \bot.^{43}$$

In words: The span of the intersection of the position and any possible momentum coordinate is equal to the origin of the vector space, which is identical to the empty set (see figure 8, right).

Figure 8: *All possible values for T span the whole space H(s) (left). The intersection of the position S with any of the momentum coordinates is equal to the origin (right). Example for three-dimensional case.*

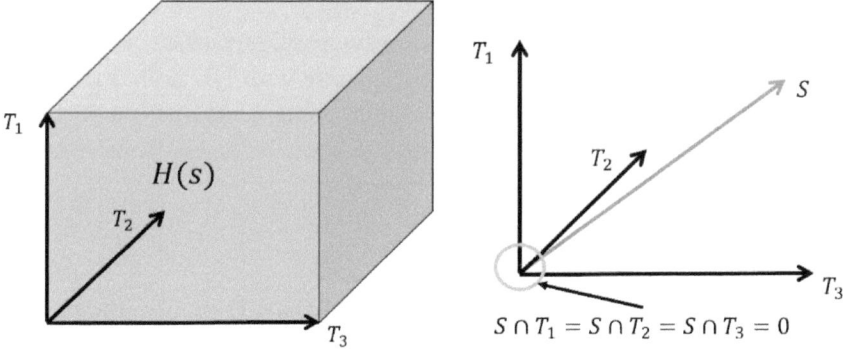

---

42 Cf. Ibid., 179.
43 Cf. Ibid., 179.

This is the reason for the failure of the distributive law in Putnam's system; the left side of the argument (position, S) is not equal to the right side (the origin, $\bot$). A formalization of the complete argument is

$S \wedge (T_1 \wedge T_2 \vee ... \vee T_R) \leftrightarrow S \not\leftrightarrow \bot \leftrightarrow \bot \vee \bot \vee ... \vee \bot \leftrightarrow (S \wedge T_1) \vee (S \wedge T_2) \vee ... \vee (S \wedge T_R)$.

Putnam's interpretation of the statements about position and momentum allows for statements of the form "The particle has a position and a momentum." to be evaluated as true, since he evaluates the disjunction of all possible values for the momentum of the particle as true in case of a given position and interprets the sentence as meaning "The particle has sharp position S and any sharp momentum.":

> "At this point let us return to the question of 'determinism' for the last time. Suppose I know a 'logically strongest factual statement' about S at t_0, and I deduce a similar statement about S after time t has elapsed – say, S_3. Then I measure *momentum*. Why can I not predict the outcome? We already said: 'because S_3 does not *imply* T_j for any j'. But a stronger statement is true: S_3 is incompatible with T_j, for all j! But it does *not* follow that S_3 is incompatible with $(T_1 \wedge T_2 \vee ... \vee T_R)$.
>
> Thus it is still true, even assuming S_3, that 'the particle has a momentum'; and if I measure I shall find it. However, S_3 cannot tell me what I shall find, because whatever I find will be incompatible with S_3 (which will no longer be true, when I find T_j). *Quantum mechanics is more deterministic than indeterministic in that all inability to predict is due to ignorance.*"[44]

Leaving open the specific momentum coordinate of the particle when knowing its position leads to a statement compatible with HU and at the same time the problem of assigning truth values to statements about a particle's position *and* momentum seems to vanish as soon as the distributive law is not considered to hold for these statements.

2.2.1 Dummett's criticism

In his article "Is logic empirical?" (1976), Michael Dummett shows that Putnam has to assume distributivity in his system because of the way he assigns meaning to disjunction:

> "It is not Putnam's opponents, but Putnam himself, who cannot, for very long at a time, 'appreciate the logic employed in quantum mechanics'; it is Putnam,

---

44 Ibid., 186–187. [Emphasis in original.]

and not they, who 'smuggles in' distributivity. For Putnam's whole argument, at this point, depends upon distributing truth over disjunction; upon assuming that, because the disjunction $B_1 \vee ... \vee B_n$ is true, therefore some one of the disjuncts must be true. And, if we assume this, it becomes impossible to see how the law which distributes conjunction over disjunction can fail to hold, since, patently, truth distributes over conjunction on any view."[45]

Thus, Dummett's criticism regards the inference from the truth of the disjunction of all momentum coordinates to the truth of one of the momentum coordinates. Since Putnam assumes this is valid, he makes the first assumption of the distributive law – a disjunction is true if one of the disjuncts is true. This is also the reason why the disjuncts can be distributed. The other necessary assumption for the distributive law cannot be rejected easily – truth has to be distributive over conjunction, since the conjunction's truth condition is that for the conjunction to be true both conjuncts have to be true. For the sake of his argument, thus, Putnam would have to reject the weaker assumption that truth distributes over disjunction. As he defined the disjunction as the span in space and not union of sets he could more easily argue for a different interpretation of its meaning or truth conditions. But this would in turn lead to his interpretation of the disjunction of all momentum coordinates to be invalid, namely the particle would not have any defined momentum. Thus, Dummett showed that the distributive law is necessarily needed for the interpretation Putnam derives from the failure of the distributive law.

2.2.2 Stachel's criticism

John Stachel argues in his paper "Do Quanta Need a New Logic?" (1986) that Putnam cannot use the classical definitions of the sentence connectives in the system he proposes and that therefore the statements which Putnam derives from the failure of the distributive law do not follow from his system. This was due to Putnam *proving* the statements with quantum-logical sentence connectives, while *interpreting* them using their classical definition:

---

45 Dummett, M. (1976): "Is Logic Empirical?" in: *Truth and Other Enigmas*, Harvard University Press, 1978, 269–289, 273.

> "This case is based, I think, on tacitly allowing the standard logical interpretation of the meaning of these propositions to hover about their discussion, while proving them by a strictly quantum-logical interpretation of their meaning."[46]

Stachel shows that applying the quantum-logical interpretation Putnam uses when proving the failure of the distributive law to statements on a particle's position and momentum leads to far weaker statements than those Putnam derives.

Stachel begins his argument citing Saul Kripke, who showed that it is impossible for the distributive law to fail when the classical interpretation of sentence connectives is applied. From this fact Stachel reasons that Putnam has to apply a non-classical interpretation in his argument[47].
According to Stachel, Kripke argues as follows: [48]

(1) He assumes that there are two magnitudes A, B and each can have the value 1 or 2.
(2) In analogy to Putnam he assumes that the statements in (a) are true while the statements in (b) are false – the disjunction of the two values that the magnitudes can have is true and the conjunction of one set value of each magnitude is false:
 (a) True: $A = 1 \vee A = 2$, $B = 1 \vee B = 2$
 (b) False: $A = 1 \wedge B = 1$, $A = 1 \wedge B = 2$, $A = 2 \wedge B = 1$, $A = 2 \wedge B = 2$
(3) Then he assumes $A = 1$. There are two possibilities for the value of B according to (1): $B = 1 \vee B = 2$.
 (3.1) He assumes $B = 1$. This leads to $A = 1 \wedge B = 1$ being true.
  → contradiction to (2b)
 (3.2) He assumes $B = 2$. This leads to $A = 1 \wedge B = 2$ being true.
  → contradiction to (2b)

Thus, assuming two magnitudes with two values in the way Putnam does leads in any case to a contradiction when the sentence connectives are interpreted classically. Stachel argues that this is why the distributive law is

---

46 Stachel, J. (1986): "Do Quanta Need a New Logic?" in: *From Quarks to Quasars: Philosophical Problems of Modern Physics*, R. G. Colodny, A. Coffa (Hrsg.), University of Pittsburgh Press, 229–347, 304.
47 Cf. Ibid., 305.
48 Cf. Ibid., 346.

valid in the classical case and if it is not valid in Putnam's logical system, then Putnam could not use the classical meaning of the sentence connectives when interpreting the statements of his system.[49]

In Putnam's system the disjunction is defined as the span of the vectors to which the propositions are assigned and all propositions described in one basis, which correspond to all possible values of a magnitude, span the whole vector space assigned to the system in question. Stachel argues that because of this definition the disjunction of all possible momentum values $\psi_1, \psi_2, ..., \psi_n$ does in fact not mean that the system has any sharp momentum but rather that a Hilbert space is assigned to the system in question which corresponds to any tautology like the one that the system exists:

> "Now, the quantum-logical interpretation of '∨' is spanning of the subspaces involved, so $\psi_1 \vee \psi_2 \vee ... \vee \psi_n$ is equivalent to the entire Hilbert space. How are we to translate this into a proposition? Presumably by 'Oscar has a Hilbert space.'[...] Perhaps we may express this more colloquially by some other version of the trivially true proposition, such as 'Oscar exists.'"[50]

Thus, Stachel shows that when rejecting the distributive law, Putnam cannot at the same time assume the classical meaning of the sentence connectives. This leads to the disjunction of all possible momentum values in conjunction with a sharp position ($S \wedge (T_1 \vee T_2 \vee ... \vee T_R)$) meaning that the system has a defined position S and not that the system has any momentum and a position at the same time.

2.3 Conclusion on the distributive law

The arguments that reject the distributive law because of a counterargument with quantum mechanical propositions seem not to work. Birkhoff and von Neumann's argument is not plausible because they do not interpret negation classically. Slightly altered, the argument still fails because the disjunction on the left side is assigned the truth value *true*,

---

49 Cf. Ibid., 305.
50 Ibid., 303. "Oscar" refers to a single-particle system introduced by Putnam in his paper 'How to think quantum-logically' in an argument corresponding to the one reconstructed in section 2.2.

although both disjuncts are *false* due to HU if the propositions involved are reformulated, as proposed by Schurz, in stronger accordance with quantum mechanics.

Putnam's argument is based on a different logical system and can therefore not be seen as directly criticising classical logic. According to Dummett, Putnam has to apply the distributive law in order to assign *true* to the disjunction of all possible momentum coordinates. This is confirmed by Stachel who argued that a classical interpretation of the sentence connectives is not possible in Putnam's system because his argument is based on a quantum-logical interpretation. The problem of the meaning of the sentence connectives in Putnam's system will be examined in section 3.1.

3. Meaning and bivalence in Putnam's quantum logical system

Putnam not only argues for the need of a new logic for quantum mechanical statements but even for refusing classical logic in favour of quantum logic which he claims to be the true logic of the world:

> "We must now ask: what is the nature of the world if the proposed interpretation of quantum mechanics is the correct one? The answer is both radical and simple. *Logic is as empirical as geometry.* [...] We live in a world with a non-classical logic."[51]

He argues that just like Euclidian geometry which had to be replaced by general relativity because of its higher empirical accuracy, classical logic might approximate reality but is not universally applicable. This is emphasized by the following relation:

$$\frac{GEOMETRY}{GENERAL\ RELATIVITY} = \frac{LOGIC}{QUANTUM\ MECHANICS}\ [52]$$

His main reason for proposing a change in logic seems to be the wish for a more accurate treatment of the – from the classical standpoint at first glance paradox – results of quantum mechanics. Putnam argues that the classical standpoint brings about inexplicable phenomena:

> "Now then, the situation in quantum mechanics may be expressed thus: we could keep classical logic, but at a very high price. Just as we have to postulate mysterious 'universal forces' if we are to keep Euclidean geometry 'come what may', so we have to postulate equally mysterious and really very similar agencies – e.g. in their indetectability, their violation of all natural causal rules, their ad hoc character – if we are to reconcile quantum mechanics with classical logic via either the 'quantum potentials' of the hidden variable theorists, or the metaphysics of Bohr."[53]

John Bell and Michael Hallett describe Putnam's motivation as follows: he is trying to save realism – a move which is required because the classical metaphysical theory (M_C) is not in accordance with the empirical data when classical physics (C) is replaced by quantum physics (Q). So Putnam

---

51 Putnam (1968), 184. [Emphasis in original.]
52 Putnam, H. (1974): "How to Think Quantum-Logically", *Synthese* 29 (1974), 55-61, 55.
53 Putnam (1968), 191.

tries to get from $Q + M_Q + L_C$ to $Q + M_C + L_Q$ (quantum physics with quantum metaphysics and classical logic to quantum physics with classical metaphysics and quantum logic), because changing logic seems to him to be less severe than giving up realism because of a counterintuitive quantum mechanical metaphysics (e. g. with hidden variables).[54]

Putnam's refusal of the distributive law raises questions. Which meaning can be assigned to the sentence connectives if the law is not valid? Putnam argues that the meaning of the sentence connectives stays the same as in classical logic, because all other classical properties of the connectives are valid in quantum logic as well. This argument is called "invariance of meaning argument" by Bell and Hallett and will be discussed together with their criticism in section 3.1.

Another question is whether or not the principle of bivalence is in accordance with Putnam's system and will be discussed in section 3.2. Putnam argues in the affirmative: as discussed above, statements about the position and momentum of a particle can be made in his interpretation of his system which would imply bivalence for quantum mechanical propositions. Bell and Hallett show that bivalence is not possible in his system because of the Kochen-Specker theorem: there is no unique mapping from Putnam's system to the set of truth values which would be needed to ensure that a truth value is assigned to at least one statement the system contains.

3.1 Invariance of meaning

Putnam argues that the distributive law (10) is not valid in quantum logic but for the sentence connectives to keep their classical meaning it was sufficient that the properties (1) – (9) hold:

"The following principles:

p implies $p \vee q$	(1)
q implies $p \vee q$	(2)
if p implies r and q implies r then $p \vee q$ implies r	(3)

all hold in quantum logic, and these seem to be like the basic properties of 'or'. Similarly

p, q together imply $p \vee q$	(4)

---

54 Cf. Bell, J., Hallett, M. (1982): "Logic, Quantum Logic and Empiricism", Philosophy of Science 49 (3), 355–379, 356.

(Moreover, $p \vee q$ is the unique proposition that is implied by every proposition that implies both p and q.)

$p \wedge q$ implies p (5)

$p \wedge q$ implies q (6)

All hold in quantum logic. And for negation we have that p and $\neg p$ never both hold $p \wedge \neg q$ is a contradiction)   (7)

$(p \wedge \neg q)$ holds   (8)

$\neg\neg p$ is equivalent to p (9)

[...]

$p \wedge (q \vee r)$ is equivalent to $p \wedge q \vee p \wedge r$ (which fails in quantum logic)   (10)

Only if it can be made out that (10) is 'part of the meaning' of 'or' and/or 'and' (which and how does one decide?) can it be maintained that quantum mechanics involves a 'change of meaning' of one or both of the connectives."[55]

Thus, Putnam considers the classical interpretation of the sentence connectives to be valid in quantum logic as long as it cannot be shown that the distributive law contributes to their meaning.

In his paper "Some conceptual problems of quantum theory" (1972), Arthur Fine (cited according to Bell and Hallett) points to the fact that a change of meaning occurs in Putnam's system just because the connectives are introduced as operations in spaces:

"It has been justly observed by Fine (1972) that this fact *already* reveals a difference in meaning between the quantum-logical and classical connectives, because in the quantum case we are concerned with a lattice of *subspaces* while in the classical case we are concerned with a lattice of *subsets*, and the operations corresponding to negation and disjunction are manifestly different in the two cases [...].".[56]

Thus, Putnam already introduces a change of meaning of the sentence connectives when choosing a structure different from a Boolean algebra, and in this case, Fine's argument is sufficient to show the failure of the invariance of meaning argument – the distributive law has not even to be considered as the change from sets to spaces already changes the meaning of disjunction and negation.

---

55 Putnam (1968), 189–190.
56 Cf. Fine, A. (1972): "Some conceptual problems of quantum theory", 3–33 (relevant here: 16–19), cited as in Bell, Hallett (1982), 362. [Emphasis in original.]

But Bell and Hallett make an even stronger case for the failure of the invariance of meaning argument. They analyse ortholattices which have been proposed as an underlying structure for quantum logic instead of Hilbert space and in fact do have the same core axioms as Boolean algebras. This is why it could be possible to define both classical *and* quantum connectives in terms of those core axioms and in this case Fine's argument could no longer be applied and quantum logic could be considered to be a more general logic as Putnam argued.

Lattices are in general algebraic structures in which the laws of associativity, commutativity and adsorption are valid and Boolean algebras are distributive lattices[57]. Ortholattices are defined as follows: "An ortholattice is a complemented bounded lattice (not necessarily distributive) satisfying the De Morgan laws and the law of double negation."[58] Just like Boolean algebras they are bounded (i.e. they have a smallest element 0 and a biggest element 1) and complemented. Further, the following theorems are valid in both structures:

"(DN) $A \leftrightarrow \neg\neg A$ Double negation
(DM∧) $\neg(A \wedge B) \leftrightarrow (\neg A \vee \neg B)$ DeMorgan-law 1
(DM∨) $\neg(A \vee B) \leftrightarrow (\neg A \wedge \neg B)$ DeMorgan-law 2"[59]

Since Boolean algebras have all the properties that ortholattices have, all Boolean algebras are ortholattices, but not the other way round (in other words: Boolean algebras are a subset of ortholattices). Thus, ortholattices are more general structures and if quantum-logical connectives could be defined in ortholattices equivalently to the classical connectives, then classical logic would be a (distributive) subset of quantum logic and Putnam's analogy would be valid. [60]

Bell and Hallett show in their paper "Logic, Quantum Logic and Empiricism" (1982) that the meaning of negation changes in any case if the connectives are defined in an ortholattice instead of Boolean algebras. Thus,

---

57 Cf. Bronstein, I, Musiol, G., Mühlig, H., Semendjajew, K. A. (2005): *Taschenbuch der Mathematik* (6. Auflage), Deutsch, Frankfurt am Main, 355.
58 Padmanabhan, R., Rudeanu, S. (2008): *Axioms for Lattices and Boolean Algebras*, World Scientific, 81.
59 Schurz, G. (2013a), 101.
60 Cf. Bell, Hallett (1982), 363.

it is not a slight change of meaning to define the connectives in a non-distributive structure:

> "We argue here that the invariance of meaning claim is false. Indeed, we show that the meaning of the negation operation in (abstract) quantum logic is quite different from the meaning of classical negation. This is sufficient to destroy the 'marginal change' contention, and thus the suggestion that quantum logic is automatically a satisfactory replacement for classical logic."[61]

They formally argue as follows: a change of meaning of a term t occurs if t can be defined with regard to entities $a, b, \ldots$ in one structure L but not in another structure L' which contains the same entities $a, b, \ldots$[62]. The structure of the argument is shown in figure 9 (1). Applied to the quantum-logical case, their argument is as follows: while all connectives can be defined using the order relation $\leq$ in Boolean algebras, this cannot be done in ortholattices where another relation is needed, "most naturally"[63] the orthogonality relation $\perp$ (see figure 9 (2)). This fact will be examined more closely in the following.

Figure 9: Bell and Hallett's argument in general (1) and with regard to quantum logic (2). (BA=Boolean algebra, OL=ortholattice).

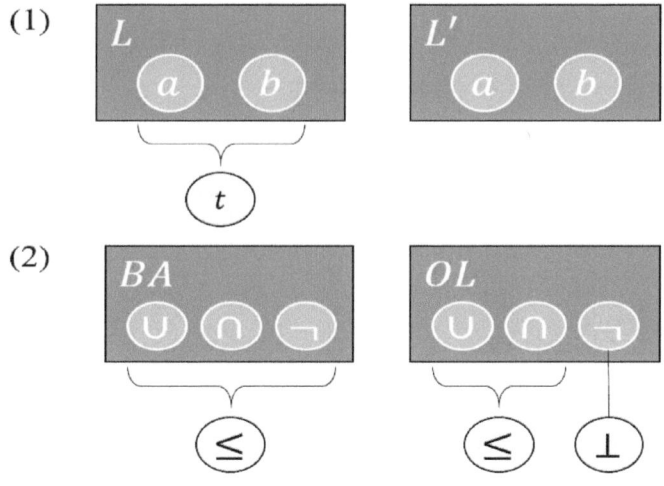

---

61 Ibid., 362.
62 Cf. Ibid., 363.
63 Ibid., 365.

Following Bell and Hallett, the sentence connectives can be defined with the order relation as follows:[64]

- Conjunction: $a \wedge b = c \leftrightarrow \{x : x \leq c\} = \{x : x \leq a\} \cap \{x : x \leq b\}$
- Disjunction: $a \vee b = c \leftrightarrow \{x : c \leq x\} = \{x : a \leq x\} \cap \{x : b \leq x\}$
- Negation: $a = b' \leftrightarrow \{x : x \leq a\} = \{x : b \wedge x = 0\}$

In words if the order relation is interpreted as implication: the conjunction of a and b is true for all x which imply both a and b. The disjunction of a and b is true for all x which are implied by a and b. a is the negation of b for all x which imply a resp. which are false in conjunction with b. It is exactly the latter definition which cannot be applied in non-distributive ortholattices and according to Bell and Hallett the most natural definition here would use the orthogonality relation:

$a = b' \leftrightarrow \{x : x \leq a\} = \{x : x \perp b\}$.[65]

In words: a is the negation of b (b') for all x which imply a resp. which are orthogonal to b.

This is why according to Bell and Hallet's definition the meaning of negation changes if it is defined in ortholattices. Thus, the same entity (negation) has different definitions in different structures which leads to their rejection of Putnam's invariance of meaning claim:

> "We may conclude from this that Putnam's invariance thesis fails even in this abstract setting, for the apparently minor 'failure of distributivity' of quantum logic is inextricably bound up with a shift in the meaning of negation. The fact that quantum negation satisfies the rules (8) and (9) mentioned by Putnam does not prevent its meaning from differing from that of its classical counterpart."[66]

3.2 Bivalence

Putnam's paper is based on the metaphysical assumption of value definiteness which means that the exact values of a particle's momentum and position do exist and just cannot be known both at the same time. This is why he considers quantum mechanics to be rather deterministic than

---

64 Ibid., 364–365.
65 Ibid., 365.
66 Ibid., 366.

indeterministic – he regards the impossibility of predicting both the momentum *and* the position coordinate's value as grounded in the impossibility to *know* them which is why, he argues, the state descriptions in quantum mechanics cannot answer all physically important questions, but they are still complete. As discussed above a particle *has* a sharp position and momentum at each point in time according to Putnam.

If the magnitudes of a quantum mechanical system are considered to have sharp values at each time which just cannot be known all at once, then statements about them would have a defined truth value which is why the principle of bivalence could be applied to statements about those properties.

Bell and Hallett argue against this assumption using the Kochen-Specker-theorem as proof which states that in the lattice of subspaces of three-dimensional Euclidian space there are no weak homomorphisms[67]. In order to apply this theorem to Putnam's case, they first introduce the assumptions (*) and (**) which are valid in Boolean algebras and to keep bivalence they have to be valid in quantum logic as well:

> "(*) for any distinct elements a, b of a Boolean algebra B there is a 2-valued homomorphism h on B such that $h(a) \neq h(b)$.
>
> Here a 2-valued homomorphism is a homomorphism to the Boolean algebra $2 = \{0, 1\}$, which, of course, can be construed as the truth value algebra, $\{T, F\}$. The statement (*) implies (and indeed, for Boolean algebras, is equivalent to)
>
> (**) for any Boolean algebra B and any element $a \neq 0$ in B, there is a 2-valued homomorphism h such that $h(a) = 1$."[68]

Thus, for the principle of bivalence to be valid there has to be a unique mapping between two algebraic structures which keeps the structure unchanged (homomorphism)[69], more specifically between the set of propositions a, b, ... and the set of truth values $\{0, 1\}$ (see figure 10) and if a proposition a is not false ($a \neq 0$), h has to assign the truth value true to it ($h(a) = 1$).

---

67 Ibid., 369.
68 Ibid., 367. [Emphasis in original.]
69 Cf. Bronstein, Musiol, Mühlig, Semendjajew (2005), 312.

Figure 10: Demonstration of the homomorphism between Boolean algebras containing the elements a and b and the Boolean algebra 2 = {0,1}.

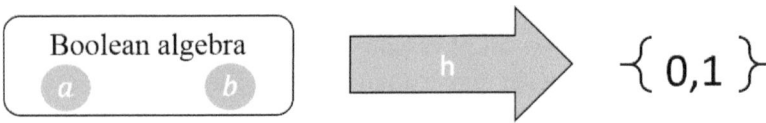

But, they continue, ortholattices fail to comply with (**). This is for quantum logic due to the existence of incompatible propositions, which are those that are true when considered separately but false if they are connected with the conjunction connective[70] (e.g. statements about specific position and momentum coordinates at the same time – $(S \wedge T_i)$. Truth values can only be assigned to the Boolean subalgebras which do not contain incompatible propositions, "but bivalence about the world concerns simultaneous assignment of truth values"[71], i.e. the principle of bivalence could only hold if the mapping h was valid for *all* statements made in quantum logic and would be in accordance to classical-logical laws for compatible propositions. The assignment of propositions to the lattice of closed subspaces of Hilbert space is done according to table 5:

Table 5: Assignment of quantum mechanical statements to the lattice of closed subspaces of Hilbert space[72]

Types of quantum mechanical propositions	Corresponding elements in the lattice of closed subspaces L_ω of Hilbert space
Compatible propositions	Boolean subalgebras of L_ω
Atomic propositions like "The value of the magnitude is such-and-such."	Lines in L_ω

A truth value must be assigned to both types of quantum mechanical propositions to keep bivalence and this assignment must be in accordance with classical-logical laws for the compatible propositions:

> "[…] given an orthocomplemented lattice L, is there a map h (representing, if you like, a simultaneous valuation) which is a weak homomorphism of L to 2,

---

70 Cf. Bell, Hallett (1982), 368.
71 Ibid., 368.
72 Ibid., 369.

that is to say, such that the restriction of h to any Boolean subalgebra of L is a homomorphism? The celebrated results of Kochen and Specker (1967) show, in effect, that for the relevant quantum lattices there can be no such map."[73]

This assumption is illustrated in figure 11: the incompatible propositions do not form a Boolean subalgebra – as the compatible propositions do – but are represented by lines in space. For a weak homomorphism, a homomorphism from the Boolean algebra of compatible propositions to the Boolean algebra (the special Boolean algebra which only contains the elements 0 and 1) would be sufficient, but Kochen and Specker showed that weak homomorphisms do not exist in 3-dimensional Euclidean space.

Figure 11: The mapping h which is needed to keep bivalence in quantum logic does not exist in ortholattices.

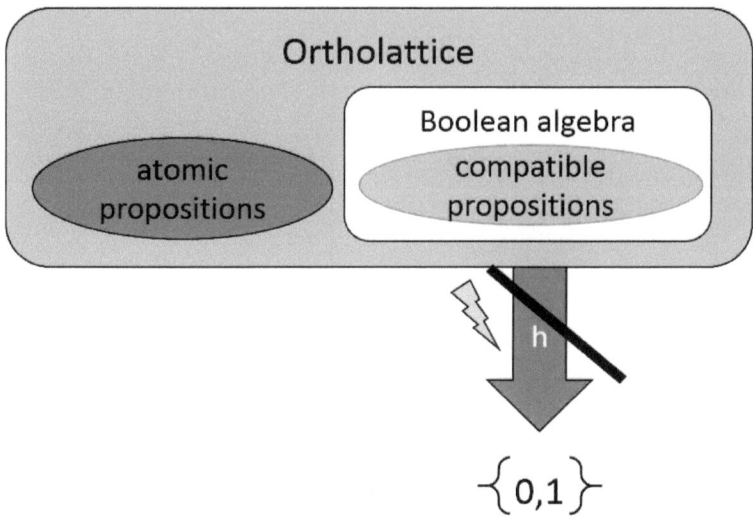

This is why it is not possible to talk about the truth or falsity of any proposition in ortholattices. Thus, the principle of bivalence is not valid in ortholattices and therefore cannot hold in Putnam's system even if his structure is changed to the more abstract one.

---

73 Ibid., 369. [Emphasis in original.]

4. Conclusion

The question whether or not logic can be empirically founded or motivated cannot be answered with the arguments discussed in this text. However, assuming that empirical data can defeat logical truths, this text points to the following results.

The counter-arguments for the distributive law are not convincing. Birkhoff and von Neumann simply do not use the classical meaning of negation in their set of propositions which makes their argument no defeater for classical logic. The distributive law was shown to be valid when the propositions were corrected in accordance with classical logic, as Popper proposed. Schurz showed that the assignment of propositions is wrong in the slightly altered form of the argument. Again, the distributive law was shown to be valid if the propositions were corrected. This means for the cases considered here that the distributive law is valid for quantum mechanical propositions if they are formulated in accordance with quantum mechanics and if the sentence connectives are interpreted in accordance with classical logic. The counter-arguments even show a higher accuracy in describing wave-particle duality and HU.

The quantum-logical system proposed by Putnam is not convincing as a replacement for classical logic as well. The distributive law is not valid in the system he proposes but his interpretation based on its failure is not possible. Dummett and Stachel showed that a non-classical interpretation of the sentence connectives is used to show the failure of the distributive law, while he interprets the result based on the distributive law resp. their classical interpretation. Therefore the statements he derives – especially on value definiteness – do not follow from his system. He either has to assume distributivity or change his interpretation. His invariance of meaning claim fails, as the proposed system is not isomorphic to Boolean algebras which leads to the definition of the negation being changed in any case (Hilbert space or ortholattice), as Bell and Hallett showed. Bivalence cannot hold in Putnam's system, since no weak homomorphisms exist in three-dimensional Euclidean space (Kochen-Specker-theorem). Those would be needed for a mapping from Boolean subalgebras in the ortholattice of all propositions to the truth values.

All in all, the discussed arguments for refusing classical logic when dealing with quantum phenomena are not plausible, as they do not reflect classical logic accurately and give no accurate account of the quantum mechanical phenomena they discuss. Putnam's quantum logic cannot be regarded a more general logic (which he takes it to be), since the meaning of the sentence connectives he introduces is not clear and the principle of bivalence is not compatible with his approach. The counter-arguments lead to the assumption that classical logic is suitable for dealing with propositions about quantum objects.

References

Auletta, G., Fortunato, M., Parisi, G. (2009): *Quantum Mechanics*, Cambridge University Press.

Bell, J., Hallett, M. (1982): "Logic, Quantum Logic and Empiricism", *Philosophy of Science* 49 (3), 355–379.

Birkhoff, G., von Neumann, J. (1936): "The Logic of Quantum Mechanics", *Annals of Mathematics* 37, 823–843.

Blackburn, S. (2008): *The Oxford Dictionary of Philosophy*, Second Edition revised, Oxford University Press, New York.

Bronstein, I., Musiol, G., Mühlig, H., Semendjajew, K. A. (2005): *Taschenbuch der Mathematik*, 6. Auflage, Deutsch, Frankfurt am Main.

Dummett, M. (1976): "Is Logic Empirical?" in: *Truth and Other Enigmas*, Harvard University Press, 1978, 269–289.

Fine, A. (1972): "Some conceptual problems of quantum theory", in: *Paradigms and Paradoxes*, R. G. Colodny (ed.), University of Pittsburgh Press, 3–33.

Feynman, R. (2010): *The Feynman Lectures on Physics*, Volume 3: Quantum Mechanics, Basic Books.

Giedra, H. (2014): *Proof System for Logic of Correlated Knowledge*. Available at: http://www.mii.lt/files/mii_dis_san_2014_giedra.pdf.

Padmanabhan, R., Rudeanu, S. (2008): *Axioms for Lattices and Boolean Algebras*, World Scientific, Singapore.

Popper, K. (1968): "Birkhoff and von Neumann's Interpretation of Quantum Mechanics", *Nature* 219, 682–685.

Putnam, H. (1968): "Is logic empirical?", *Boston Studies in the Philosophy of Science 5*, Dordrecht, Holland: Reidel, 216–41. Page numbers in the text are taken from the reprint titled "The Logic of Quantum Mechanics" in Putnam (1979): *Philosophical Papers* Vol. 1: 174–197.

Putnam, H. (1974): "How to Think Quantum-Logically", *Synthese* 29, 55–61.

Schurz, G. (2011): *Einführung in die Wissenschaftstheorie*, WBG, Darmstadt.

Schurz, G. (2013a): *Logik I: Einführung in die Aussagen- und Prädikatenlogik*. Available at: https://www.phil-fak.uni-duesseldorf.de/fileadmin/Redaktion/Institute/Philosophie/Theoretische_Philosophie/Schurz/scripts/SkriptVLLogikINeu.pdf.

Schurz, G. (2014): *Logik II*. Available at: https://www.phil-fak.uni-duesseldorf.de/fileadmin/Redaktion/Institute/Philosophie/Theoretische_Philosophie/Schurz/teaching_materials/VLLogikII2014.doc.

Schurz, G. (2013b): *Philosophy of Science, A Unified Approach*, Routledge, New York.

Shankar, R. (1994): *Principles of Quantum Mechanics* (2nd edition), Plenum Press, New York.

Stachel, J. (1986): "Do Quanta Need a New Logic?" in: *From Quarks to Quasars: Philosophical Problems of Modern Physics*, R. G. Colodny, A. Coffa (eds.), University of Pittsburgh Press, 229–347.

Straumann, N. (2013): *Quantenmechanik: Ein Grundkurs über nichtrelativistische Quantentheorie*, 2. Auflage, Springer Spektrum.

Zettili, N. (2001): *Quantum Mechanics, Concepts and Applications,* John Wiley & Sons.

Joint Conclusion
(by Jan Philipp Dapprich and Annika Schuster)

One of the central themes in this book can be summarized as follows: Is quantum mechanics compatible with both 'classical logic' and 'classical ontology'? A negative answer to this question would force us to choose between classical logic and classical ontology. Putnam argues that in the light of quantum mechanics classical logic is wrong and has to be replaced by a quantum logic. He wishes to save realism, including value definiteness. According to him, this would save us from inacceptable ontological implications which quantum mechanics when applied to classical logic would lead to, namely that it is incompatible with realism and value definiteness. Annika Schuster showed in the second part of the book that on the one hand Putnam's quantum logic can itself not fulfil these needs (since the meaning of the connectives in his system is not clear and his system is not compatible with bivalence) and secondly that classical logic is not necessarily incompatible with quantum mechanical propositions as long as these are formulated respecting the value indefiniteness of quantum mechanics.

Whether logic can be empirically motivated is not decidable from the arguments discussed. But it is possible to state that even if it was in fact empirical, the discussed arguments are unsuccessful in establishing an incompatibility between the empirical phenomena and classical logic.

Does this mean we have to reject realism? In the first part of this book Philipp Dapprich argued that we indeed have to choose between classical logic and value definiteness. However, a rejection of value definiteness doesn't imply a rejection of realism altogether. Even if some observables only have values when measured, other quantities, such as the wave state, may well be interpreted in a realist fashion.

Quantum mechanics is revolutionary in many ways. Besides making value definiteness impossible, it also precludes the possibility of deterministic prediction. But claims that it is in conflict with core aspects of standard scientific methodology are exaggerated. It is not the method of science that needs to be revised, but rather our understanding of quantum mechanics.

We were unable to decide conclusively whether or not such a revision of our understanding should include a revision of the very fundamentals of quantum mechanics, as Penrose suggests. The least we can say is that, in the light of the problems arising when attempting to combine quantum mechanics with general relativity, the choice of an interpretation of quantum mechanics may well be relevant. Choice of interpretation can, however, not solely be based on our ontological background, as most interpretations are compatible with a wide range of ontological positions.

Philosophische Grundlagen der Wissenschaften und ihrer Anwendungen
Philosophical Foundations of the Sciences and Their Applications

Herausgegeben von / Edited by Gerhard Schurz

Vol./Bd.	1	Jens Paulßen: Das Problem der Kurvenanpassung. Das Balancieren der Ansprüche der Einfachheit und der Genauigkeit. 2011.
Vol./Bd.	2	Matthias Rosner: Theorien zur Gestaltung der modernen Organisation. Eine interdisziplinäre Perspektive. 2011.
Vol./Bd.	3	Alexander Christian: Wissenschaft und Pseudowissenschaft. Ein Beitrag zum Demarkationsproblem. 2013.
Vol./Bd.	4	Karim Baraghith: Kulturelle Evolution und die Rolle von Memen. Ein Mehrebenenmodell. 2015.
Vol./Bd.	5	Jan Philipp Dapprich / Annika Schuster: Philosophy and Logic of Quantum Physics. An Investigation of the Metaphysical and Logical Implications of Quantum Physics. 2016.

www.peterlang.com

www.ingramcontent.com/pod-product-compliance
Ingram Content Group UK Ltd.
Pitfield, Milton Keynes, MK11 3LW, UK
UKHW041438190426
11946UKWH00021B/16